Eyre Champion De Crespingny

A New London Flora

Or Hand Book to the Botanical Localities of the Metropolitan Districts

Eyre Champion De Crespingny

A New London Flora
Or Hand Book to the Botanical Localities of the Metropolitan Districts

ISBN/EAN: 9783337128050

Printed in Europe, USA, Canada, Australia, Japan

Cover: Foto ©berggeist007 / pixelio.de

A

NEW LONDON FLORA;

OR,

Handbook to the Botanical Localities

OF THE

METROPOLITAN DISTRICTS.

COMPILED FROM THE LATEST AUTHORITIES, AND FROM PERSONAL OBSERVATION.

BY

EYRE CH. DE CRESPIGNY, M.D., M.R.C.S.,

LATE OF H.M. MEDICAL SERVICE IN INDIA; FORMERLY CIVIL SURGEON IN THE SOUTHERN CONCAN; AND AT ONE TIME ACTING CONSERVATOR OF FORESTS AND SUPERINTENDENT OF BOTANICAL GARDENS IN THE BOMBAY PRESIDENCY.

LONDON:

HARDWICKE AND BOGUE, 192, PICCADILLY.

1877.

LONDON:

PRINTED BY WILLIAM CLOWES AND SONS,

STAMFORD STREET AND CHARING CROSS.

A HUMBLE CULTIVATOR

OF THE SAME TOO FREQUENTLY FALLOW FIELD

Dedicates,

WITH FEELINGS OF THE GREATEST RESPECT AND ESTEEM,

THIS SMALL RESULT OF HIS LABOURS

TO

THE AUTHOR OF THE 'CYBELE BRITANNICA';

THAN WHOM NO ONE HAS DONE MORE FOR TOPOGRAPHICAL AND GEOGRAPHICAL BOTANY,

EITHER IN PAST OR IN PRESENT TIMES.

PREFACE.

A New London Flora is much required. The author of the present publication has undertaken the task in the hope that it may be of use to students of practical botany, and has based it, as it should be, not only upon material supplied by libraries, but also upon actual collections in the field. The only handbook of the kind still extant, he believes, is Mr. Daniel Cooper's 'Flora Metropolitana;' but this is now out of date, and with its numerous references to the 'Botanist's Guide' of 1805, and other ancient authority, is unadapted to the requirements of the day. Moreover the arrangement of the work is confused, the nomenclature often obsolete, and there is no index to the species. It is a well-known fact that vegetation everywhere alters to a certain extent with the lapse of time; it does so to a marked degree in the neighbourhood of London, where the disturbing elements of clearing, draining, building, enclosing, are perpetually at work. Consequently, very many of the suburban localities indicated by Mr. Cooper have long disappeared, together with the plants which grew there; and what were once rural solitudes are now "portions and parcels" of this vast metropolis. The neighbourhood of Putney is still tolerably productive, so also is Hampstead Heath; but there is little out of the common to be obtained in Epping Forest nearer than Loughton. Blackheath is a bare common; Greenwich marshes have been drained and converted into market gardens; those below Woolwich, into pasturage; a small portion of Harrow Weald is still unenclosed; likewise, Stanmore Heath; otherwise, most of the heaths and commons round about exist only in name. On the other hand, the facilities of locomotion afforded by railway communication are now so great, that students can readily extend their researches to the more distant sandy heathlands and chalk ranges of Kent and Surrey, to the woodlands and cornfields of Essex and Herts, as well as to the banks of the Thames above Richmond and below Erith.

In addition to an alphabetical list of all flowering plants and Cryptogams, down to Mosses and their allies inclusive, growing within an average radius of thirty miles from London, with their respective records of where found, as far as it is practicable not to overlook them, and omitting all such as are in any way doubtful by reason of antiquity of date—a series of localities is appended, with lists of the plants to be found there, preceded by short topographical descriptions, most of them having been subjected more or less to personal investigation, and all arranged to suit the requirements of half-holiday makers and students, with whom time is a consideration. Lichens and Fungi are also included, but only the more generally distributed and important species. The lists have also been made out with reference, less to what *has been*, than to what does actually exist and in appreciable quantity. The nomenclature is that adopted by Mr. Watson in his 'Catalogue of British Plants,' latest edition, 1874 (Messrs. Hardwicke and Bogue, Piccadilly).

With regard to these personal investigations, the author begs to state that they were carried on for seven years consecutively under circumstances of impaired health, contracted in India, which compelled him to take much exercise in the open air of the country, whenever the state of the weather would admit of it; and having for many years previously devoted much of his leisure time to botanical pursuits, the opportunity was taken advantage of to obtain the required information; an expenditure of time however, which might have been deemed by others, better qualified for the task than himself, not only wearisome but unprofitable. Every plant gathered or seen *in situ* by the author is marked with an asterisk, but only where the observation was duly recorded at the time; vague recollections of having seen them elsewhere are untrustworthy. Collectors will observe that new localities for rare plants are few and far between. In this direction the author's researches have been far from satisfactory; in fact, it is a question now-a-days, not what there is, but what there is not, in the way of rarity. Those who wish for rarities must go far to find them. The causes before referred to, will account for the disappearance of many of them ; the extermination of others has been often ascribed to the indiscriminate rapacity of collectors themselves, and the growing passion for collecting. However, for purposes of study one plant is as good as another; all that is further necessary is sufficient variety, and many a humble weed is possessed of structural peculiarities as instructive and interesting as any to be found in plants of less frequent occurrence. Those who would know the geographical range of the vegetation round London will find in every locality the characteristic plants of the same, duly recorded ; but to the mere collector the constant repetition of the same thing in one part of the country and of another thing in another part will appear tedious and superfluous. Given a series of plants, and the nature of the locality which produced them can be determined at a glance. In addition to the series common to certain localities, many of them possess some speciality or other peculiar to themselves: the black Bog Rush grows on Bagshot Heath ; the marsh Cinquefoil, on Elstead Common ; the ivy-leaved Campanula (Wahlenbergia) grows at Witley Lagg ; the Bog Gentian on Chobham Common, and so forth : The Fritillary is at home in the basin of the Thames ; the white Beak Rush is rare, except on the sandy heaths of S.W. Surrey. Many plants are of uncertain appearance in their respective localities ; especially such as are designated colonists. They appear and disappear, according to the nature of the crops, &c. Excess of heat, of moisture, of drought, or the contrary may exercise an influence in one way or the other ; persons therefore, visiting localities in search of any plant in particular, must not count too surely upon success in finding it.

One difficulty with which the general student may have to contend is the nomenclature adopted in the Catalogue of 1874, with which this publication is in accord. Of 237 species additional, nearly all are promotions from the rank of varieties. The total for all Britain, exclusive of Charads and minor Cryptogams, amounts to 1665, of which number no less than 1250 are computed to occur within the average thirty-mile radius assigned to this compilation; reports, however, as to the actual occurrence, generally, of a percentage of these plants, are considered by Mr. Watson doubtful.[1] As this radius is almost entirely restricted to province iii. of the 'Cybele' and Compendium, the separate list of these aggregates, with their respective segregates, which is given, is drawn up with reference thereto exclusively.

It can, after all, matter little whether, once named, these segregates be regarded as mere varieties, or as distinct species; every botanist is at liberty to form his own conclusion on the subject, but the chief objection lies "in the inextricable confusion caused by the transfer of names from one variety to another, and adoption of obsolete ones to replace those till lately in use; so that, in several cases, it is hard to say whether the aggregate plant is meant by the term, or one of its segregates" (Cybele Comp.). The author of the 'Cybele Britannica,' who has devoted forty years to the study of the geographical distribution of plants, justly observes, with regard to this system of species splitting, that "empirical names are given to species, much after the methods resorted to by florists in naming their varieties of Roses and Geraniums. *It is individuals, not species, that are described technically.* The thing is overdone; species are subdivided on differences so slight that descriptive botanical language cannot make them intelligible, without figures or specimens. There is no constant distinction made between species and varieties, other than a decision by individual opinion in each case; the splitters leave others to prove the contrary; hence conflict of opinion; and decision may be altered by succeeding generations." (Ibid.) Whether we endorse Mr. Watson's opinions or not, it must surely be apparent to every one, that in attaching equal value to the distinctions between species and varieties we necessarily lessen the value of those between species and species; and are led on to consider this question, what then, after all, constitutes a species and what a variety? One glance at the Catalogue will reveal, in addition to the promotions just made, a

[1] See Appendix to Index.

formidable array of intermediate forms waiting, it may be, their turn to rise from obscurity. In fact, it comes to this, "that species and varieties, the latter especially, are optional and arbitrary, as artificial arrangements of dried specimens and of portraits of individual plants: any one having equal right to make either more or fewer species and varieties of the same materials. We are bewildered by their ill-defined pettiness, and the tendency is to make book botany attractive to those only who are incapable of large and extended views." (Ibid.) Besides, what proofs have we of the permanency of these subspecies?

As a natural consequence of this system, many records of localities for plants in the aggregate can no longer be relied on, because it is impossible to determine now, to which of its segregate forms the record may have referred; and only in the most recent Floras and periodicals are localities for any of these segregates to be found; moreover, "the records of locality reporters are often unreliable by reason of deficient knowledge, carelessness in observation, inaccuracy of language, and one-sided statements." (Ibid.)

The nomenclature of the Moss tribe is that of Mr. Berkeley's 'Handbook of the British Mosses,' the adoption notwithstanding by the Reverend Professor of the monoicous and dioicous theory of foreign botanists, and the questionable arrangement consequently of the genera based upon this fallacy, and an over-estimated importance attached to the areolation of the leaves. But, for descriptive particulars of these little plants, and for the excellence of their illustrations, the student can consult no better authority. The author has elsewhere expressed his opinions on the disputed doctrine of the fertilisation of flowerless plants; the subject is immaterial from a practical point of view, and no purpose would be answered by any remarks in this place with regard to it. Those who are interested in Scale-mosses will find all the new names in their proper places, although the nomenclature according to Hooker has been retained for particulars; an advantage of arrangement which admits of their being grouped together and thus readily distinguished from the Mosses proper, &c., with which they are associated in the section. The genus Jungermannia was split into nine genera by Gray, and subsequently increased to forty-eight![1] In Cryptogams, as well as in flowering plants, the number of species has been exaggerated; far so, indeed, beyond the requirements of science.

A list of authorities consulted in this compilation is added to the other lists; it has not been deemed advisable to look into any of earlier date than the 'New Botanist's Guide' of 1834. Actual occurrence in the present, and not the local history of plants in the past, is what has been aimed at, however imperfectly that object may have been attained.

For lists of Aliens, Casuals, Extinctions, and other excluded species, the reader is referred to the 'London Catalogue.'

[1] See remarks, Endlicher, 'Genera Plantarum.'

CONTENTS.

	PAGE
PREFACE	v
LIST OF AUTHORITIES CONSULTED	xiv
LIST OF LOCALITIES	xv
LIST OF AGGREGATES, SEGREGATES, AND SYNONYMS	xviii
I. FLOWERING PLANTS	1
II. CRYPTOGAMS	79
SUPPLEMENT TO INDEX	99
LOCALITIES	103

LIST OF AUTHORITIES CONSULTED.

'Annals of Natural History.'
Berkeley: 'British Mosses,' &c.
Babington: 'Manual.'
Bentham: 'Handbook.'
Boswell-Syme: 'English Botany.'
Brewer, J. A., and Salmon, 'Flora of Surrey.'
Brewer: 'New Flora of Reigate.'
Britten: 'Contributions to a Flora of Berkshire.'
'Botanical Gazette,' Henfrey Ed.
Cooper, D.: 'Flora Metropolitana,' 1837.
Gibson, G.: 'Flora of Essex.'
Hooker and Arnott: 'British Flora.'
Hooker: 'Journal of Botany,' &c. (various).
Irvine, 'London Flora,' 1838.
Jardine: 'Magazine of Zoology and Botany,' 1837–38.
Jenner, 'Flora of Tunbridge Wells.'
'Journal of the Quekett Microscopical Society.'
Luxford: 'Flora of Reigate,' 1838.
Melvill: 'Flora of Harrow,' 1864.
'Magazine of Natural History.'
'Naturalist.'
'Phytologist.'
Trimen and Dyer: 'Flora of Middlesex.'
Watson, H. C.: 'Cybele Britannica,' and Compendium to ditto.
,, ,, 'Topographical Botany.'
,, ,, 'New Botanist's Guide.'
Webb and Coleman: 'Flora of Herts,' with Supplement to ditto.
Wilson: 'Bryologia.'
Smith, Curtis, Balfour, and others: 'Botanist's Chronicle,' and other periodicals; 'Science Gossip' (for Moss localities). W. G. Smith (for Fungi); M. C. Cooke (for Hepaticæ); Reports of Bot. Exchange and Bot. Locality Record Clubs.

LIST OF LOCALITIES.

		PAGE
1.	Hampstead Heath	103
2.	Barnes Common	104
3.	Banks of the Thames from Putney to Kew	105
4.	Putney Heath and Wimbledon Common	106
5.	Lanes and Roadsides about Hendon, Neasdon and Kingsbury	107
6.	Willesden: banks of the Brent, and Paddington Canal	108
7.	Pastures and lanes about Tottenham and Edmonton; banks of the Lea, and Lea Canal	109
8.	Epping Forest, and copses, lanes, &c., about Chingford, Woodford and Walthamstow	110
9.	Blackheath, and the marshes below Greenwich	111
10.	Charlton Wood and chalk-pit; Woolwich sandpits; Shooter's Hill	112
11.	Wandsworth Common	113
12.	Roadsides, copses, and waste places about Norwood	114
13.	Mitcham Common	115
14.	Banks of the Thames, with bordering ditch and meadows between Kew and Kingston; Richmond Hill and Ham Common	116
15.	Roadsides about Isleworth, Twickenham, Teddington, Hounslow, and banks of the Cran	117
16.	Pinner and Oxhey Woods; meadows about Pinner and Ruislip; Ruislip reservoir	117
17.	Harrow Weald Common; Stanmore Heath; Elstree reservoir	118
18.	Totteridge Green and Hadley Common	119
19.	Epping Upper Forest	120
20.	Hainault Forest and banks of the Roding	122
21.	Marshes between Woolwich, Plumstead, Erith, and opposite shore	122
22.	Plumstead Common; Bosthall Heath; Abbey Wood; Erith sand-pits	123

	PAGE
23. Chislehurst Common	124
24. Hayes and Keston Commons	125
25. Shirley Common and the Addington Hills	126
26. Croham Hurst, and the adjoining fields, banks and roadsides	127
27. Purley Downs and Riddlesdown; Smitham Bottom	128
28. Farden (or Farthing) Downs, with Coulsdon and Kenley Common	129
29. Sutton and Banstead Downs, with bordering fields	131
30. Ditton Marsh, and Esher Commons	132
31. Moulsey Hurst, and banks of the Mole about Moulsey and Esher	134
32. Banks of the Colne, between Uxbridge and Harefield	135
33. Colney Heath	137
34. Broxbourne and Wormley Woods	138
35. Warley Common	139
36. Purfleet	140
37. Greenhithe and Dartford	141
38. Darne (or Darent) Wood	143
39. North Downs, near Sevenoaks	143
40. Reigate Hill and the Wray Common	145
41. Reigate Heath, Redhill and Earlswood Commons	146
42. Merstham, and hills east of Merstham; Redstone Hill	147
43. The Betchworth Hills	148
44. About Buckland and Brockham	150
45. Ranmer Common, hills west of Dorking and the White Downs	150
46. Guildford, and the hills east of Guildford, including Shiere and Albury	152
47. Whitemoor Common	154
48. Bisley Common, Pirbright Heath, and Cow Moor	154
49. Woking Heath; Horsell Common, and banks of the Basingstoke Canal	155
50. Weybridge, St. George's Hill, and banks of the Thames about Walton	156
51. Chobham Common	157
52. Bagshot Heath	158
53. The Hog's Back	159
54. Puttenham, Elstead and Crookesbury Commons	159
55. Frensham Common	161
56. Witley and Thursley Commons	161
57. Godalming	162
58. Leith Hill and the Holmwood	162

	PAGE
59. Tilgate Forest	164
60. Felbridge	164
61. High Rocks and Waterdown Forest, near Tunbridge Wells	165
62. Hills east of Wrotham	165
63. Cobham and Cuxton	166
64. About Northfleet and Gravesend	167
65. Grays and Tilbury	168
66. Southend and Canvey Island	169
67. Norton Heath, Ongar and Fyfield	170
68. Essex cornfields	171
69. Hertford Heath and surrounding woods	171
70. The Lea Valley about Hatfield, Hertford and Ware	172
71. Upper Colne district	174
72. Tring and Aldbury	175
73. Gerard's Cross and Stoke Commons	177
74. Burnham Beeches and Farnham Common, with adjoining woods	177
75. Thames district, above and about Windsor	178

LIST OF AGGREGATES, SEGREGATES, AND SYNONYMS.

AGGREGATES.	SEGREGATES.	SYNONYMS.
Ranunculus (aquatilis)	circinatus.	
	fluitans.	
	peltatus.	
	diversifolius.	
	Drouettii.	pantotrix.
	trichophyllus.	heterophyllus.
	Baudotii.	
	intermedius.	tripartitus.
Fumaria (capreolata)	pallidiflora,	
	and var.	
	confusa.	
	muralis.	
,, densiflora	. .	micrantha.
,, (tenuisecta)	parviflora.	
,,	Vaillantii.	
Diplotaxis tenuifolia	. .	Sinapis ten.
,, muralis	. .	,, mur.
Brassica (polymorpha)	Napus.	
	,,	campestris.
	Rutabaga.	Napus.
	Rapa.	
Sisymbrium Alliaria	. .	Erysimum Alliaria.
Cardamine (hirsuta)	hirsuta.	
	sylvatica.	
Arabis perfoliata	. .	Turritis glabra.
Barbarea (vulgaris)	vulgaris.	eu-vulgaris.
	arcuata.	
	stricta.	
	intermedia.	
Cochlearia (polymorpha)	officinalis.	
	danica.	
Alyssum maritimum	. .	Koniga maritima.
Senebiera didyma	. .	Coronopus didyma.
,, Coronopus	. .	,, Ruellii.
Viola (hirta)	hirta.	

AGGREGATES.	SEGREGATES.	SYNONYMS.
Viola (hirta)	permixta.	sepincola (var.)
„ (canina)	sylvatica.	
	canina.	
	arenaria.	
	lactea.	
	stagnina.	
„ (tricolor)	tricolor	eu-tricolor.
	arvensis.	var. of tricolor.
	Curtisii.	
	lutea.	
Polygala **vulgaris**	vulgaris.	
	oxyptera.	
	depressa.	
Cerastium **pumilum**	. .	? triviale.
Stellaria **aquatica**	. .	Cerastium aquaticum (i.e. Malachium).
Alsine tenuifolia	. .	Arenaria ten.
Sagina apetala	maritima.	
	apetala.	
	ciliata.	
Spergularia (marina)	neglecta.	
	marginata.	
	rupestris.	
Hypericum (quadrangulum)	tetrapterum.	
	dubium.	
	bæticum.	
Ulex (nanus)	Gallii.	
	nanus.	eu-nanus.
Medicago **falcata**	sylvestris.	
	falcata.	
Trifolium incarnatum	. .	Trifolium Molinerii var. incarnatum.
Lotus (corniculatus)	corniculatus.	
	tenuis.	
„ (angustissimus)	angustissimus.	
	hispidus.	
Vicia (sativa)	sativa.	
	angustifolia.	
Prunus (communis)	spinosa.	
	insititia.	
	domestica.	
Agrimonia Eupatoria	Eupatoria.	
	odorata.	
Potentilla Tormentilla	Tormentilla.	
	procumbens.	
Rubus (fruticosus)		

In lieu of the six species of Arnott, exclusive of *R. cæsius*, we have forty segregates according to Professor Babington's views (3, i.e. province 3, of the 'Cybele' and the home counties).

AGGREGATES.	SEGREGATES.	SYNONYMS.
[Species according to Hooker and Arnott:	suberectus 3.	umbrosus.
	fissus.	
suberectus	plicatus 3.	
rhamnifolius	affinis 3.	lentiginosus.
fruticosus	Lindleianus 3.	nitidus.
carpinifolius	rhamnifolius 3.	
glandulosus	incurvatus 3.	
corylifolius]	imbricatus.	
	latifolius.	
	discolor 3.	fruticosus.
	thyrsoideus 3.	argenteus.
	leucostachys 3.	
	Grabowskii.	
	Colemani.	
	Salteri.	calvatus.
	carpinifolius 3.	
	villicaulis 3.	sylvaticus.
	macrophyllus 3.	Schlechtendalii amplificatus.
	mucronulatus.	
	Sprengelii 3.	Borreri.
	Bloxamii.	
	Hystrix 3.	
	rosaceus 3.	pallidus.
	pygmæus 3.	
	scaber 3.	Babingtonii.
	rudis 3.	Leightonii.
	Radula 3.	Lingua, Bab.
	Kœhleri 3.	
	fusco-ater.	Schleicheri, Colemanni.
	diversifolius 3.	
	Lejeunii 3.	
	pyramidalis.	
	Guntheri 3.	
	humifusus 3.	
	foliosus.	hirtus, Menkii.
	glandulosus 3.	Bellardi, dentatus, rotundifolius, fuscus, rosaceus.
	Balfourianus 3.	tenui-armatus.
	corylifolius 3.	
	althæifolius 3.	
	tuberculatus 3.	

AGGREGATES.	SEGREGATES.	SYNONYMS.
Geum (rivale)	intermedium. rivale.	
Rosa (spinosissima)	spinosissima. rubella.	
„ involuta	. .	Sabini, var. involuta.
„ (rubiginosa)	rubiginosa. sepium.	
„ stylosa	. .	Systyla.
Pyrus (Aria)	Aria, pinnatifida, and three others not mentioned.	
Epilobium (tetragonum)	tetragonum. obscurum.	
Callitriche verna	verna. obtusangula. stagnalis.	
„ autumnalis	hamulata. truncata. autumnalis.	pedunculata.
Chærophyllum Anthriscus	. .	Anthriscus vulgaris.
„ sativum	. .	„ Cerefolium.
„ sylvestre	. .	„ sylvestris.
Scabiosa arvensis	. .	Knautia arvensis.
Silybum Marianum	. .	Carduus Marianum.
Carduus crispus	. .	acanthoides (agg. olim).
Arctium Lappa et Bardana	majus. minus. intermedium. nemorosum.	
Matricaria Parthenium	. .	Chrysanthemum Parthenium.
„ inodora	. .	„ inodorum.
„ Chamomilla	. .	„ Chamomilla.
Tanacetum vulgare	. .	„ Tanacetum.
Filago germanica	germanica. apiculata. spathulata.	
Senecio paludosus	paludosus. palustris.	
„ campestris	. .	Cineraria campestris.
Crepis fœtida	. .	Barkhausia fœtida.
„ taraxacifolia	. .	„ taraxacifolia.

Hieracium—

In lieu of eighteen species for all Britain, given in the sixth edition of the Catalogue, we have a total of thirty-five. In so far as they relate to the flora of the London districts, six only of this number occur—

Hieracium Pilosella.

AGGREGATES.	SEGREGATES.	SYNONYMS.
Hieracium murorum.		
„ vulgatum.		sylvaticum.
„ tridentatum.		
„ umbellatum.		
„ boreale.		
Specularia hybrida	. .	Campanula hybrida.
Wahlenbergia hederacea	. .	„ hederacea.
Erythræa (Centaurium)	Centaurium, pulchella, and two others not mentioned.	
Limnanthemum nymphæoides		Villarsia nymphæoides.
Scrophularia (aquatica)	Balbisii.	
	Ehrharti.	
Veronica spicata	spicata.	
	hybrida.	
„ serpyllifolia	serpyllifolia.	
	humifusa.	
Orobanche major	. .	Orobanche Rapum.
(minor)	Picridis.	
	hederæ.	
	minor.	
Mentha (rotundifolia)	rotundifolia.	
	alopecuroides.	
„ (aquatica)	pubescens.	
	citrata.	
	hirsuta.	
„ (sativa)	sativa.	
	rubra.	
	gracilis.	
	pratensis.	
	gentilis.	
Thymus (Serpyllum)	Serpyllum.	
	Chamædrys.	
Calamintha (officinalis)	menthifolia.	
	sylvatica.	
Lamium (amplexicaule)	amplexicaule	
	intermedium.	
Anchusa arvensis	. .	Lycopsis arvensis.
Utricularia (vulgaris)	vulgaris.	
	neglecta.	
Anagallis (arvensis)	arvensis.	
	cærulea.	
Plantago lacustris	. .	Plantago Coronopus.
Atriplex portulacoides	. .	Obione portulacoides.
„ (patula)	angustifolia.	
	erecta.	

AGGREGATES.	SEGREGATES.	SYNONYMS.
Atriplex deltoidea	. .	Atriplex hastata.
Rumex Hydrolapathum	Hydrolapathum.	
" "	maximus.	
Parietaria diffusa		Parietaria officinalis.
Sparganium affine et minimum	. .	Sparganium natans.
Potamogeton natans	natans, polygonifolius and another.	
" heterophyllus	heterophyllus and another.	Potamogeton Proteus.
" gramineus	acutifolius.	
	obtusifolius.	gramineus.
	zosterifolius.	
" pusillus	. .	Potamogeton compressus.
Zannichellia (palustris)	palustris.	
	pedicellata.	
	polycarpa.	
Orchis militaris	militaris.	
	purpurea.	Orchis fusca.
	Simia.	
" latifolia	latifolia.	
	incarnata.	
Gymnadenia conopsea	. .	Orchis conopsea.
" albida	. .	Habenaria albida.
Habenaria (bifolia)	bifolia.	
	chlorantha.	
Epipactis latifolia	latifolia, and two others.	
Scilla nutans	. .	Hyacinthus non-scriptus.
Smilacina bifolia	. .	Maianthemum bifolium.
Juncus (communis)	effusus	
	diffusus.	
	conglomeratus.	
Carex disticha	. .	Carex intermedia.
" flava	flava.	
	Œderi.	
Echinochloa Crus-galli	. .	Panicum Crus-galli.
Digraphis arundinacea	. .	Phalaris arundinacea.
Agrostis Spica-venti	. .	Apera Spica-venti.
Psamma arenaria	. .	Ammophila arundinacea.
Calamagrostis Epigejos	. .	Arundo Calamagrostis.
" lanceolata	. .	" Epigejos.
Phragmites communis	. .	" Phragmites.
Avena elatior	. .	Arrhenantherum avenaceum.

AGGREGATES.	SEGREGATES.	SYNONYMS.
Festuca (Myurus)	ambigua.	
" "	pseudomyurus.	
	sciuroides.	
" rubra	. .	Festuca duriuscula and var.
Bromus racemosus	racemosus.	
	commutatus.	
Triticum junceum	junceum.	
	acutum.	
	pungens.	
Lomaria Spicant	. .	Blechnum boreale.
Aspidium aculeatum and angulare	. .	Polystichum aculeatum and angulare.
Nephrodium Filix-mas, and others	. .	Lastræa Filix-mas, and others.

N.B. The Synonyms have special reference to the sixth edition of the 'London Catalogue.'

A NEW LONDON FLORA.

I. FLOWERING PLANTS.

* *Seen or gathered by author* in situ. (?) *Doubtful whether still to be found there.* (E. de C.) *On his authority.*

ACER CAMPESTRE.—Woods and hedges, frequent, 5–6.

Acer Pseudo-platanus.—Plantations, 5–6.

ACERAS ANTHROPOPHORA.—Chalky pastures, and old chalk-pits, now scarce, 6. Greenhithe and Northfleet chalk-pits, abundant in the former (Saturday Half-holiday Guide, 1874); Box Hill and Betchworth Hills; Purley Downs and downs about Coulsdon; *olim*, chalk-steppes beyond Tring; Reigate Hill;* chalk-pits on the Hog's Back, and old chalk-pits near Harefield (?); downs, about Albury and Shiere; Knockholt and Wrotham.

ACHILLEA MILLEFOLIUM.—Pastures and waysides, frequent, 6–9.

ACHILLEA PTARMICA.—Damp pastures and wet places on heaths and waste ground, frequent, 7–8. (Plenty on Ditton Common.)

ACORUS CALAMUS.—Ponds and marshy places near rivers, not frequent, 6. Barnes Common;* marshy flat near Walton Bridge; ponds on Totteridge Green;* Roding River between Woodford and Chigwell; Colne River, between Uxbridge and Harefield; Coulsdon; Bisham; between Richmond Park and Wimbledon, in several places; pond at Bayfordbury; Staines Common; by the Brent at Greenford.

ACTINOCARPUS DAMASONIUM.—Pools and ditches on gravelly commons, frequent formerly, now rarely met with, 6–7. Headley and Walton Heaths; Earlswood Common (?); Totteridge Green (?); Tooting and Leatherhead Commons; pond on Ditton Marsh, sparingly;* and pond below Winter Downs, Esher;* Mitcham Common, in a small pool* (E. de C.); ponds near Ilford; Uxbridge Moor, towards Denham (?); Epping Forest, near Woodford (?); Putney Heath (?); Shalford Common (?); Warley Common; Holmwood Common; Felbridge pools, perhaps; Winkfield Plain; and at Bracknell, near Windsor; Waterdown Forest.

ADONIS AUTUMNALIS.—Casually in cornfields, 5–7, uncertain. About Dartford, Greenhithe, and Crayford, formerly; Croydon; Acton; Reigate; near Warley Common; between Cobham and Cuxton.

ADOXA MOSCHATELLINA.—Hedgebanks and shady places in woods, 4–5. Woods about Brentwood; Fyfield; copse near Chingford Hatch,* copse between Boreham Wood stat. and Stanmore Heath;* Croydon towards Selsdon;* woods about Coulsdon, and about Harefield; also about Watford, and elsewhere in Herts; hedgebank in a field between Aldbury and Aldbury Nowers Wood.*

ÆGOPODIUM PODAGRARIA.—Wet places and ditches, generally near houses and gardens, frequent, 5.

ÆTHUSA CYNAPIUM.—Fields and gardens, common, 7–8.

AGRIMONIA EUPATORIA.—Fields, waste places, and roadsides, common, 6–7.

AGRIMONIA ODORATA.—Among osiers east end of Telegraph Wood, Claygate.

AGROSTIS ALBA.—Pastures, roadsides, &c., common, 7–8.

AGROSTIS CANINA.—Moist heaths and moory places, common, 6–7. (Plenty, lower part of Hampstead Heath.)

AGROSTIS SETACEA.—Local. Plenty on Bagshot Heath (left of the road from Chobham); also in Waterdown Forest, on one spot near Heathfield; also on Bisley Common.

AGROSTIS SPICA-VENTI.—Cornfields, not common, but plenty where it occurs, 6–7. About Croydon, Charlton, Coulsdon; between Ilford and Barking; Esher;* Northfleet, and Gravesend; Guildford, towards Farnham; Cheshunt.

AGROSTIS VULGARIS.—Meadows, pastures, and banks, everywhere, 6–7.

AIRA CÆSPITOSA.—Moist, shady places and borders of fields, common, 6–7. (Plenty, lower part of Hampstead Heath.)

AIRA CARYOPHYLLEA.—Gravelly hills and pastures, frequent, 6–7. (Hampstead Heath;* Buckland Hill.)

AIRA FLEXUOSA.—Heaths and hilly places, common, 7. Hampstead Heath, sandy hillocks towards the 'Spaniards;'* Warley Common;* Harrow Weald Common.

AIRA PRÆCOX.—Sandy hills and pastures, frequent, 5–6. (Hampstead Heath.)

AJUGA CHAMÆPITYS.—Cultivated slopes and plateaux on the Kentish and Surrey chalk range, local but frequent, 4–10. Fields of Cobham Park;* Dartford; Croydon (near Croham Hurst*); fields bordering Epsom and Banstead Downs;* Box Hill;* Buckland Hill; hills W. of Dorking* (E. de C.), near Purfleet.

AJUGA REPTANS.—Moist pastures and woods, common, 5–6.

ALCHEMILLA ARVENSIS.—Pastures and banks on gravelly soil, frequent, 5–8. Barnes Common;* Chislehurst Common;* Dartford Heath,* &c.

ALCHEMILLA VULGARIS.—Hilly pastures, rare; also in low meadows; Stanmore (?); between Bushey and Watford (?); and bank of Bourne Hall moat; Gorhambury Park, and in an old lane leading thence to Gorhambury; Rickmansworth Common Moor;* Totteridge Park; damp meadows by the banks of the Mole, near Dorking; and near Flanchford Bridge; Great Canfield, near High Roding; woods, Tring, and wood W. of the monument, Aldbury; Hill Park, near Westerham; by brook in a pasture between Brickendonbury, and lane to the green; Hertford Heath, by the road to Essendon, opposite Roxford; low pasture between Essendon and West-end; also in a valley not far from Essendon Place; Great Berkhampstead, by the footpath from Gossoms End to Magdalene Chapel; meadows between Stanboro' and Hatfield Woodfall; field between

Ruislip reservoir and road to Harefield; hilly pastures about Stanmore; Parkhurst near Abinger.

ALISMA PLANTAGO.—Margins of ponds, rivers, and ditches, frequent, 7.

ALISMA RANUNCULOIDES.—Ditches and turfy bogs, uncommon, 7–8. Epping Forest; pond, Breakspeares; Warley Common (?); Putney Heath;* banks of the canal above and about Woking* (E. de C.), and in turfy pools near the canal* (E. de C.).

ALLIUM OLERACEUM.—Rare, near the sea, 7. About Rochester (?); wall at Milford; hedgebank, Cold Harbour Lane Croydon; in cornfields at Welwyn, Herts.

ALLIUM URSINUM.—Moist woods and hedgebanks, especially ditch banks, 5–6. Lane between Chingford and Walthamstow;* banks of the Mole, between Southgate and Barnet; banks of the Brent below Totteridge;* lane leading from Welsh Harp to Kingsbury;* banks of the rivulet, near Chislehurst station.*

ALLIUM VINEALE.—Fields and waste places, rare, 6. Thames bank, near Teddington;* meadow below the mill, by the Mole at Esher;* Fyfield; Southend; near Thorley; and near the gravel pits on Moulsey Hurst; hill above Vale Cottage, Albury; chalk-pit at the bottom of Church street, Guildford; dry knolls in meadows between Hertford and Ware; field between Welwyn road and S. part of Thieves lane; fields between Hoddesdon and the Rye House, and between Hoddesdon and Haileybury; lanes about Sunbury and Laleham.

ALNUS GLUTINOSA.—Watery places, and by rivers, frequent, 3.

ALOPECURUS AGRESTIS.—Cultivated fields, and occasionally roadsides, frequent; plenty in cornfields on a clay soil (Tottenham and Edmonton), &c.,* 5–10.

ALOPECURUS GENICULATUS.—In pools, especially pools in gravel pits, and on a gravelly soil, 5–8.

ALOPECURUS FULVUS.—In pools, and in similar situations to the above, but less frequent, 7. Gravel pits in Epping Forest, near Woodford * (E. de C.); gravel pit near Tottenham, N. of the church* (E. de C.).

ALOPECURUS PRATENSIS.—Meadows and pastures, common, 5–6.

ALOPECURUS BULBOSUS.—Rare, local. Salt-marshes about Northfleet (?) 7.

ALSINE TENUIFOLIA.—Sandy fields, rare (?). About Coulsdon; on sand; roadside, Tottenham.*

ALTHÆA OFFICINALIS.—Near the sea; rare, 8–9. From Tilbury to Southend.

Alyssum calycinum.—Incidental, rare. Fields near Epping; Hitchin Common; near Hoddesdon; plenty by the New River.

Alyssum maritimum.—A garden waif; near houses occasionally. Foot of Redhill;* Purfleet (?).

AMMOPHILA.—*See* PSAMMA.

ANAGALLIS ARVENSIS.—Cornfields and fallow fields, frequent, 5–11.

ANAGALLIS CÆRULEA.—Cornfields in the chalk districts, rare. Fields between Sutton and Banstead Downs (?); cornfields, slope, and summit of Box Hill; also of the Betchworth Hills; cornfield below Buckland Hill,*[1] in company with *A. arvensis;* between Keston and Down, about Broomfield; and on Warley Common (?).

[1] Perhaps a variety only of *A. arvensis.*

ANCHUSA ARVENSIS.—Cornfields and waste places, frequent, 6–7 (plenty in a cornfield going from Ditton Marsh to Telegraph Hill)* (E. de C.).

Anchusa sempervirens.—Waste grounds about old ruins and buildings, very rare. Near Rochester, formerly.

ANAGALLIS TENELLA.—Bogs, especially on peaty commons, in company with Sphagnum, 7–8. Bog near High Beech, Epping Forest;* Esher Common;* Reigate Heath;* Leith Hill;* Shirley Common (?); frequent in peaty bogs, &c., on the Surrey heaths;* Hoddesdon Marsh; Bell Bar Bog.

Anemone apennina.—Rare, 4. In Wimbledon Park; and in woods about Shiere and Guildford; private grounds between Mitcham and Sutton.

ANEMONE NEMOROSA.—Woods and coppices, frequent, 4–5 (on Hampstead Heath; Epping Forest, near Woodford, plenty).

ANEMONE PULSATILLA.—Rare. Downs near Aldbury Nowers Wood;* downs about Streatley, and banks S.E. of Ravensburg Castle, borders of Herts, 5.

ANGELICA SYLVESTRIS.—Ditches, moist woods, and marshy places near rivers, common, (ditches by the Lea, plenty), 7.

ANTHEMIS ARVENSIS.—Cornfields, uncommon; more frequent on the chalk, 6–7. Great Warley; Tilbury; Epping, in fields near; about Hertford; open field between Hoddesdon and the Rye House; gravelly ground between Hatfield and Holwell; and field on Hertford road, near N.E. boundary of Hatfield Park; about Teddington.

ANTHEMIS COTULA.—Waste places, and by roadsides, not uncommon, 6–9 (cornfields about Tottenham and Edmonton).

ANTHEMIS NOBILIS.—Gravelly pastures, frequent. Hampstead Heath; Wimbledon Common; Ham Common, &c.,* 8–9.

ANTHOXANTHUM ODORATUM.—Fields and pastures; common, 5–6.

ANTHRISCUS.—*See* CHÆROPHYLLUM.

ANTHYLLIS VULNERARIA.—Everywhere very frequent in the chalk districts; Hertford; St. Albans; Tring; about Croydon and Sanderstead;* Greenhithe, and road from Dartford to Darent Wood;* Reigate Hill;* Box Hill;* and all along the range of the North Downs from Cuxton to Farnham.*

Antirrhinum majus; old walls in several places about London, 7–8. Plenty on the railway chalk cutting near Sutton station.*

ANTIRRHINUM ORONTIUM.—Dry cornfields not very common, 7–10. Haslemere, near Godalming; Brockham; cornfield between Ditton marsh and Telegraph Hill;* plentiful in a cornfield near Chobham, coming from Chertsey* (E. de C.); near Croydon.

APERA.—*See* AGROSTIS SPICA-VENTI.

APIUM GRAVEOLENS.—Ditches by the Thames, on both sides of the river, common (plenty in the Plumstead Marshes; and between Greenwich and Woolwich);* 8–9.

AQUILEGIA VULGARIS.—Woods and coppices, rare, 5–7. Pastures about Harefield *olim* (?); Box Hill;* (foot of and in Westhumble Lane); woods at Coulsdon; Tring woods;* Netley Wood; Shiere, copses on the chalk near Dorking; and at Hightrees farm; Powis Wood, Godalming; Norwood.

ARABIS THALIANA.—Old walls, frequent, 4–5.

ARABIS HIRSUTA.—Banks in the chalk districts, 6. Smitham Bottom, near Croydon;* bank bordering Farthing Downs* (E. de C.); Banstead Downs; Box Hill and Betchworth Hill; Reigate Hill; Mickleham; Tring; (bottom of Aston Hill); near Shiere; Ashdown copse; Ranmore Common; pits, Guildford.

ARABIS PERFOLIATA.—Hedgebanks and roadsides, not common. Frequent N.W. of Hertford; copse, hillside beyond Golding's Wood; gravel field E. side of Hatfield; between Hatfield Park and Cole Green, plentiful; roadside between Hatfield and St. Albans; by the Park, Claremont *olim* (?); about Shiere and Albury; sandy lanes, Frensham; about Denham; between Waltham and High Beech; roadside by Cottenham Park, Wimbledon; Godalming, in coppices; Cookham wood; near Brocket Hall, Herts.

ARCTIUM MAJUS.—Damp woods and waste places; not uncommon; Wormley wood;* Warley woods; Hatfield.

ARCTIUM MINUS.—Roadsides and waste places; frequent, 7–8.

ARCTIUM NEMOROSUM.—Rare; Welwyn.

ARENARIA SERPYLLIFOLIA.—Walls and dry places; frequent, 6–7. Walls about Croydon; fields between Sutton and Banstead.* Var. *leptoclados*, Hatfield.

ARENARIA TENUIFOLIA.—*See* ALSINE.

ARENARIA TRINERVIS.—Shady woods and moist places, 5–6. Charlton Wood; Harrow Weald; borders of Darent Wood, towards Greenhithe, plenty.*

ARMERIA MARITIMA.—Muddy salt-marshes; rare near London, 4–9. In a field above Northfleet; between Leigh and Southend;* Purfleet; Tilbury.

Armoracia rusticana.—Frequent near gardens and habitations in wet places. (Plenty near West-end railway station on the railway bank towards Finchley road.)

ARNOSERIS PUSILLA.—Sandy fields; rare, 6–7. About Weybridge *olim*, not now (?); field near Chobham, coming from Chertsey (?); strawberry beds, Bexley Heath (?); sandy field near Esher and Oxshott; banks near Farnham, Frensham, and in fields on the flanks of Chobham ridges; in a gravelly field behind the public-house at the Hammer ponds, Thursley Common; Petersham in a sandpit; Hampton Court Park, formerly plentiful, and in fields adjoining; about Teddington; field left of road from Hersham green to St. George's Hill, less than a mile from the former; field by a fir wood, right of road to Milford from Witley station.

ARRHENATHERUM AVENACEUM.—*See* AVENA ELATIOR.

ARTEMISIA VULGARIS.—Hedges and waste places, common, 8. (Plenty along the banks of the canal, Tottenham.)

ARTEMISIA ABSINTHIUM.—Dry banks and waste places, generally near houses and gardens. Chalk-pits, Greenhithe and Northfleet, *olim* (?); Hampstead Heath, 2 plants;* Purfleet; White Roding; near Wellington College, plentiful; Stanstead, by the road to Hertford; by the roadside between Watford and St. Albans, plenty; Great Berkhampstead, on waste ground.

ARTEMISIA MARITIMA.—Salt-marshes by the Thames; (plentiful about Erith, Plumstead, &c.*), 8.

ARUM MACULATUM.—Damp hedgebanks, and copses, common; (plenty in the lanes about Neasdon).

ARUNDO.—*See* CALAMAGROSTIS and PHRAGMITES.

ASPARAGUS OFFICINALIS.—Thames bank below Woolwich; rare, 6–8. Near Erith, a plant or two.*

Asperugo procumbens.—On rubbish in waste places, very rare; 6–7.

ASPERULA CYNANCHICA.—Banks in the chalk districts, 6–7. There common all along the range from Cuxton to the Hog's Back;* roads from Croydon to Selsdon and Sanderstend;* Banstead Downs;* from Dartford to Darent Wood;* banks about Tring;* chalk-pit, Hertford.

ASPERULA ODORATA.—Cool and damp woods, abundant where occurring, 5–6. Darent Wood;* Charlton Wood (?); woods on Reigate and Buckland hills;* Tring woods;* Burnham Beeches;* woods on the flank of Leith Hill;* wood near Harefield,* &c.

ASTER TRIPOLIUM.—Muddy banks of the Thames, plenty both sides below Greenwich;* in profusion also at Canvey Island;* and on the banks of the Medway at Cuxton,* 9–10.

ASTRAGALUS GLYCYPHYLLOS.—Woods and thickets in a chalky soil, not common, 6. Woods about Harefield; Greenhithe and Darent;* Coulsdon; borders of Cobham Park;* wood at Purfleet (?); Hatfield Heath; border of Warwick Wood; Frith Hill, Godalming; old chalk-pit on the Hog's Back; chalk hills E. of Merstham; Headley lane.

ATRIPLEX ANGUSTIFOLIA.—Waste, and cultivated ground, frequent, 7–10.

ATRIPLEX HASTATA.—Waste, and cultivated ground, not uncommon, 6–10.

ATRIPLEX DELTOIDEA.—*See* A. HASTATA.

ATRIPLEX ERECTA.—In similar situations.

ATRIPLEX BABINGTONII.—Muddy shores of the Thames, towards Leigh;* not common.

ATRIPLEX LITTORALIS.—Muddy shores of the Thames, frequent towards Erith;* Northfleet; Purfleet;* and beyond, on both sides of the river,* 7–9.

ATRIPLEX PORTULACOIDES.—Shores of the Thames, between Leigh and Southend,* where sandy; plenty there, 7–8.

ATRIPLEX ARENARIA.—Thames shore towards Southend; in sandy places,* not common, 7–9.

ATROPA BELLADONNA.—Copses in the chalk districts, hedges, and waste places, 6–8, rare. About Dorking and Box Hill;* between Merstham and Godstone; base of the chalk hills near Quarry Farm; chalk hills about Kemsing; Moor Park; plentiful formerly about Tring; Coulsdon; Cobham (?); foot of the Betchworth Hills; Norbury Park; woods about Denbies; Ranmore Common; Epping Forest (?); W. Clandon; Old Park Wood, Harefield.

AVENA FATUA.—Cornfields, frequent, in Essex especially; between Hatfield, Broad Oak, and Fyfield, frequent;* Claygate.

AVENA FLAVESCENS.—Dry meadows, and pastures, common, 7.

AVENA ELATIOR.—Hedges, and ditch sides, common; (plenty about London, in hedges by the high-roads), 6–7.

AVENA PRATENSIS.—Pastures, heathy, and hilly places, 6–7. Gravel Hill, Northfleet; marshes between Greenwich and Woolwich (?); Charlton Wood (?); foot of the hills opposite Brockham.* (Localities confused with that of *A. pubescens:* Mr. Watson.)

AVENA PUBESCENS.—Dry pastures in the chalk districts, not unfrequent, 6–7. Banstead Downs,* scarce; meadows by the Thames, Moulsey; foot of chalk hills W. of Dorking;* meadows by the Mole, about Esher (from seeds brought by floods from the Downs?); also E. of Merstham;* and E. of Shoreham.*

Avena strigosa.—Cornfields (rare), 6–7. Farnham; occasionally between the Rye House and Hoddesdon.

BALLOTA NIGRA.—Hedgerows, and waste places, especially near habitations (plenty on the hedges about the outskirts of London), 6–10.

BARBAREA VULGARIS.—Hedges and ditches, frequent, 5–8. By the Thames, both sides of Hammersmith Bridge, Surrey side;* by the ditch on Barnes Common.*

BARBAREA ARCUATA; B. INTERMEDIA; B. STRICTA.—These forms, now considered as distinct species, may be met with in similar situations. Between Hertford and Ware; Broxbourne Wood; brooks at Cheshunt; brook at Totteridge; Roxeth.

Barbarea præcox.—Waste places, rare. Purley Downs (?); road from Dorking to the Holmwood; between Betchworth and Wonham Mill; near Brockham; banks of the Wey, near Guildford; Ruislip; Pinner;* wastes about Thames Ditton; gravelly field near N.E. boundary of Hatfield Park on Hertford road; Twickenham; Apperton, near the canal bridge, plenty.

BARKHAUSIA.—*See* CREPIS FŒTIDA and C. TARAXIFOLIA.

BARTSIA ODONTITES.—Fields and roadsides, frequent, 6–8. (Plenty along the banks of the Paddington Canal.)

BELLIS PERENNIS.—Pastures, frequent, 2–10.

BERBERIS VULGARIS.—Woods, hedges, and plantations, 5–6. Hedges about Northfleet, and from Dartford Heath to Greenstreet Green, roadside right; between Tring stat. and Aldbury; field by Lea Bridge road; Lea districts; Iver.

BETA MARITIMA.—Muddy banks of the Thames, both sides of the river, frequent; between Plumstead and Erith;* about Northfleet,* Gravesend,* Rochester, and between Leigh and Southend,* 6–9.

BETULA ALBA.—Woods, frequent, 4–5. Epping Forest;* Burnham Beeches;* woods about Leith Hill;* Warley Common;* and on Chislehurst Common.*

BIDENS CERNUA.—Sides of ditches, ponds, and streams, common; (plentiful about Tottenham and Edmonton in ditches by the Lea), 7–10.

BIDENS TRIPARTITA.—In similar situations, and in marshy places, frequent; (ditches by the Lea, and banks of the Roding, Colne, and on Colney Heath, abundant), 7–9.

BLYSMUS COMPRESSUS.—Boggy pastures near springs, very rare. Shirley Common, *olim* (?); wet meadow near Bagshot Heath (Winch MSS., and N. B. G.); bog left of Redhill road, 2 miles from Merstham (?); about Ham Pond below Redstone Hill (possibly); and bogs, foot of Cockshott Hill; boggy pasture, now enclosed, opposite Beddington Park gate, Mitcham Common;* in Hatfield Forest, (probably by the large pond in the park). In the rill near Dulwich Wells, *olim*.

Borago officinalis.—Waste places near habitations, a garden waif, incidentally.

BRACHYPODIUM SYLVATICUM.—Woods, and hedges, frequent, 7. Plenty in the shady lanes about London;* Epping Forest.*

BRACHYPODIUM PINNATUM.—On the open downs of the chalk range, frequent, 7. Plentiful on Mickleham Downs;* Reigate;* Dorking;* Merrow Downs,* &c.

Brassica Napus.—Cornfields, and waste ground, frequent; a waif of cultivation, 5–6. *B. Rutabaga*, a mere variety.

BRASSICA RAPA.—In similar situations, frequent; (banks of the New River,*) 4–7.

BRIZA MEDIA.—Meadows, and pastures, chiefly on the chalk; (plentiful on the Surrey Downs, everywhere *), 6.

BROMUS ASPER.—Moist woods, and hedges, common; (plentiful in lanes on outskirts of London *), 6–7.

BROMUS ERECTUS.—Sandy fields, and roadsides, chiefly in the chalk districts, 6–7. Plentiful at the foot of the downs, facing the Weald of Kent;* of Reigate Hill,* and hills E. of Merstham;* Bisham Wood, and Winter Hill, Berks; fields between High Rocks and Waterdown Forest; Tring; chalk-pit, above Harefield.

BROMUS STERILIS.—Waste ground, fields, and hedges, common; (plenty about Willesden and Brondesbury, Tottenham, &c.), 6.

BROMUS COMMUTATUS.—Cornfields, and roadsides, frequent (lanes everywhere *), 6–7.

BROMUS RACEMOSUS.—Hedgebanks, in cool and shady places (at Hendon *), 6.

BROMUS SECALINUS.—Cornfields in the chalk districts, not frequent. Field near Dartford, on the road, right, to Darent Wood * (E. de C.); near Dorking * (E. de C.); about Brasted, 6–7.

Bromus arvensis.—Cornfields, rare, 7–8. Near Keston; fallow field near Dartford.* (E. de C.—Yes; H. C. Watson.)

BROMUS GIGANTEUS.—Shady woods, and moist hedges, unfrequent (?), Charlton Wood. In a marshy place by the Thames, between Putney and Hammersmith; a few plants only * (E. de C.); 7–8.

BRYONIA DIOICA.—Thickets and hedges; common, (frequent in hedges by the roadside environs of London), 5–9.

BUNIUM FLEXUOSUM.—Woods and pastures, frequent. Hampstead Heath;* lower part, and in Highgate Wood;* Esher Common;* Pinner Wood, plenty, &c.,* Warley Common, 5–6.

BUPLEURUM FALCATUM.—Local; roadside beyond Norton Heath, plentiful.* On Reigate Heath (?); (none now).

BUPLEURUM ROTUNDIFOLIUM.—Cornfields, in a chalky or gravelly soil over chalk, elsewhere unfrequent. Cornfields on the Surrey Downs, *olim;* Epsom, Leatherhead, and Boxhill (?); between Guildford, and St. Martha's Chapel, towards the Merrow Downs (?); about Dartford, Greenhithe, and Purfleet (?). Cornfields in Essex E. of Bishop's Stortford;* about Fyfield;* in Herts W. of Bishop's Stortford, near Thorley. In the weald, near Felbridge, &c.; Streatley; near Sutton; cornfields near the Hermitage, on Buckland Hill.

BUPLEURUM TENUISSIMUM.—Muddy shores of the Thames, not frequent;

near Stroud; east of Tilbury Fort * (E. de C.). Ealing Common, *olim* ! (?) (*Sison Amomum* mistaken for it?) 8–9.

BUTOMUS UMBELLATUS.—Ditches, ponds, &c., frequent, 8–9. Ditches bordering the river Lea, and canal, plenty;* banks of the Colne,* Thames,* Roding,* &c., 6–7.

Buxus sempervirens.—Chalky hills in Surrey, about Coulsdon, Boxhill, in abundance;* Harefield; Epping Forest (?); 4–6.

CAKILE MARITIMA.—Local, sandy places on the shore at Southend, 6–7.

CALAMAGROSTIS EPIGEJOS.—Damp, shady woods, scarce, 7. Old Park Wood, Harefield; and shady lane near Harefield, leading to Rickmansworth (?); Larkswood near Chingford; near Salter's buildings, Walthamstow (?); elsewhere in Epping Forest (possibly); hedgebanks about Brockham (?) Coombe Wood, Wimbledon (?); copse near Watford over mouth of the railway tunnel (?); about Virginia Water; Weston Wood, Albury; wood near High Rocks, towards Tunbridge Wells; Pryor's Wood, Hertford Heath (?); (none there now, nor on the heath itself). In Ball's Wood, near Hertford * (yes; H. C. Watson); Chigwell; up Mangrove Lane, half a mile from Hertford (not seen, but Phragmites there in plenty); hedge between Broxbourne Wood and lane, from Goose Green to Hertford; near Beulah Spa, Norwood (?).

CALAMAGROSTIS LANCEOLATA.—Ball's Wood near Hertford (?); confounded with *C. Epigejos* (?) Newland's Wood, near Rickmansworth, very rare. (This, a much smaller plant, is more frequent in the fens than elsewhere.)

CALAMINTHA ACINOS.—Cultivated and fallow fields in the chalk districts, and in gravelly soil with chalky substratum, frequent, 7. Between Sutton and Banstead;* Harefield; Box Hill;* hills, west of Dorking;* Reigate Hill;* and hills, E. of Shoreham; Tring; St. Albans.

CALAMINTHA CLINOPODIUM.—Common in the chalk districts. Epsom Downs; Box Hill;* Mickleham;* Dorking; everywhere along the range, and occasionally on gravelly soil with a chalky substratum, 7–9. (Banks of the Thames,* but washed down from chalk downs of Berks.)

CALAMINTHA MENTHIFOLIA.—Waysides, and borders of fields in the chalk districts, or near them, and on gravelly soil overlying the chalk, frequent, 7–9. Chalk-pits, Harefield (?); Guildford chalk-pits (?); hedges about Mickleham Lane, leading from Nutfield to the downs * (E. de C.). Roadside about Cuxton;* plentiful in a lane at the foot of the downs, from near Cuxton to Wrotham * (E. de C.). Lane near Chobham; near Egham; about Albury; Shiere; Norbury Park.

CALAMINTHA NEPETA.—Banks, and waysides in the chalk districts; rare, 7–8. Road from Dartford to Greenstreet Green (?); old record, lane leading to the river at Harefield * (a few plants); bank behind the railway platform, Box Hill stat.* (E. de C., a few plants); near Chelmsford and elsewhere in the neighbourhood; Grays; Great Warley; between Watford and St. Albans; Hatfield Park.

CALLITRICHE VERNA.—Ditches, and pools, frequent, 4–9. Everywhere.

CALLITRICHE STAGNALIS.—In similar situations; on Putney Heath; Frensham Pond; in ditches between Hértford and Ware; and in Broxbourne and Wormley woods.

CALLITRICHE HAMULATA (and TRUNCATA).—Shalford Common.

CALLITRICHE AUTUMNALIS (?).—A northern form.

CALLUNA VULGARIS.—Heaths, and moors, common, 6–8. Hampstead Heath, &c.* (Rare in Herts: Hertford Heath;* Wormley Wood.)*

CALTHA PALUSTRIS.—Marshes, common (plenty in marshy places by the river Lea, Thames Bank, &c.), 3–6.

Camelina sativa.—Cultivated fields, occasionally, rare, 6–7. Tilford near Farnham; near Epping; fields near Keston Common; Watford; between Stanstead and Ware. Among wheat in open upland fields between Ashtead and Leatherhead.

CAMPANULA GLOMERATA.—Dry chalky pastures, frequent in the chalkdowns, 7–8. Mickleham;* Box Hill;* in the warren, Epsom Downs;* Ranmore Common, Harefield; chalk steppes, S.W. of Tring; Dorking chalkpit; about Farnham; Hatfield Broad Oak.

CAMPANULA HYBRIDA.—*See* SPECULARIA.

CAMPANULA HEDERACEA.—*See* WAHLENBERGIA.

CAMPANULA TRACHELIUM.—Shady woods, and lanes, frequent, 7–8. Burnham Beeches; frequent in copses on the chalk range, and in lanes below and near it; woods about Gatton, and Cold Harbour;* wood near Rickmansworth; about Brockham;* Dorking;* Guildford;* and Godalming; Old Park Wood, Harefield; between Shorn and Stroud; hills, E. of Merstham;* in a lane at the foot of the downs, leading from Wrotham to Cuxton* (E. de C., plentiful), and in a copse near Walton Downs, Epsom;* Bayford woods, and copses thence to the Lea; Tring.

CAMPANULA PATULA.—Pastures and hedges, rare, about Croydon? (*olim*), 7. Sandy lanes, Frensham near Farnham; side of Chobham lane, near Windlesham church; sparingly.

CAMPANULA RAPUNCULUS.—Gravelly wastes, 7. Duppas Hill, Croydon (?) (*olim*); about Dorking and Mickleham; Dartford (?); Enfield (?); Farnham; Hersham; sparingly, *olim*, about Esher; Box Hill.

CAMPANULA RAPUNCULOIDES.—Rare between Wotton and Leith Hill.

CAMPANULA ROTUNDIFOLIA.—Heaths and roadsides in healthy countries, common (Hampstead Heath; Barnes Common; Putney Heath, &c.), 7–9.

CAPSELLA BURSA-PASTORIS.—Fields and waste places, everywhere, 5–10.

CARDAMINE AMARA.—Wet meadows near rivers and river-sides; not unfrequent, 4–6. By the Thames, in ditches between Mortlake and Kew;* about Dorking; Godalming; by the Ravensbourne at Lewisham; banks of the Colne near Harefield;* and at Uxbridge;* swamps below hills E. of Merstham; in some alder-copses near Reigate; Moor Park; Farnham; by the Cran at Babe Bridge.*

CARDAMINE PRATENSIS.—Ditches and wet meadows, common (plentiful by the Thames and by the Lea, &c.), 4–6.

CARDAMINE HIRSUTA.—Moist shady places everywhere, especially on ditch-banks, 3–8.

CARDAMINE IMPATIENS.—Rare, 5–8. Godalming woods and in Catshall copse near Godalming.

CARDUUS ACANTHOIDES.—*See* C. CRISPUS,

CARDUUS ACAULIS.—Dry gravelly and (especially) on dry chalky pastures, 6–7, there abundant. Banstead Downs;* Surrey downs everywhere;* Moulsey Hurst;* Hatfield Forest;* Kentish downs;* Tring.*

CARDUUS ARVENSIS.—Waste places, fields, commons, roadsides everywhere, frequent, 7.

CARDUUS CRISPUS.—In similar situations with the preceding, and almost as frequent, 6–8.

CARDUUS ERIOPHORUS.—Waste ground in the chalk districts, rare, 7–8. Stone chalk-pit near Greenhithe (? *olim*; often mistaken for *C. lanceolatus*, 'Cybele Britannica'); near White Roding.

CARDUUS LANCEOLATUS.—Pastures, roadsides, and waste places, common, 7–8.

Carduus Marianum.—See Silybum.

CARDUUS NUTANS.—Waste ground in dry stone and chalky soil, common in the chalk districts, 7. Boxhill;* Banstead Downs;* Tring; in fields below Albury Nowers Wood, abundant occasionally with white flowers;* Mickleham, about Northfleet and Purfleet;* fields near Dartford towards Darent Wood.*

CARDUUS TENUIFLORUS.—Waste sandy ground and in chalky fields, not common. Cornfield on Reigate Hill;* and elsewhere occasionally in the chalk district; Purfleet; Tilbury; Boxmoor, near.

CARDUUS PRATENSIS.—Wet pastures and on damp heaths and moors; not uncommon, 6–8. Ditton Common, plenty;* Esher Common;* Putney Heath;* Wimbledon Common;* Rickmansworth Common Moor; about Dorking; Croydon; Epping Forest.

CARDUUS PALUSTRIS.—Moist meadows and shady places; frequent, 7. (Plenty lower part of Hampstead Heath;* Wimbledon Common and Coombe Wood.*)

CARDUUS FORSTERI (hybrid) Surrey; (? *olim*; an obscurity; Cybele Brit.)

CAREX ACUTA.—Marshy meadows and wet pastures; 5–6, by the Lea; pasture by the Thames near Putney* (E. de C.); pond in Ball's Wood; Hoddesdon marsh; by the Thames about Ditton and Sunbury; by the Mole near Esher mills and S. of Woodhatch; Reigate; banks of the Roding towards Loughton* (E. de C.); ditches between Hertford and Ware.

CAREX AMPULLACEA.—Bogs on peaty commons and borders of ponds in similar localities. Felbridge; Reigate Heath;* Earlswood Common; Witley Lagg;* Hatfield Forest; banks of the Roding, not common, 5–6.

CAREX ARENARIA.—Sandy sea-shores, 5–6 (rarely inland). On sandy shore near Southend; sandy commons about Farnham! and Frensham!*

CAREX AXILLARIS.—Marshes, rare. The Ledds near Esher Common; Merstham pools (?).

CAREX BŒNNINGHAUSENIANA.—Swampy woods, very rare, 6. In a willow bed near Bourne mill, Farnham, ('Botanical Gazette'). Ball's Wood near Hertford, in ponds there towards S.W. corner of Hertford Heath;[1] near Reigate.

CAREX BINERVIS.—Dry heaths and moors, frequent. Wimbledon Common;* Esher Common;* abundant on the Bucks heaths; Hertford Heath; Blackfan Wood, Bayford; Harrow Weald Common; Whitemoor Common;* Pirbright Common;* Stoke Common;* Farnham Common by Burnham Beeches;* Gerard's Cross Common;* 6, Epping Forest; Warley Common.

CAREX CÆSPITOSA.—*See* C. STRICTA.

1 Doubtful: see Report for 1874, p. 88 of Botanical Locality Record Club. An unsatisfactory species; Comp. Cybele.

Carex curta.—Bogs and ditches, 6. Reigate Heath;* Whitemoor Common; lanes in a hollow near Neasdon;* hedge at Totteridge;* Virginia Water.

Carex dioica.—Bogs, rare, 5–6; perhaps often overlooked. Shirley Common; peat bogs in Waterdown Forest.

Carex divisa.—Marshy meadows and borders of ditches near the sea; plentiful in the ditches and pastures below Woolwich and beyond; on both sides of the river.*

Carex disticha.—Marshy ground, 5–6, not common, 6. In the ravine, Wimbledon Common;* woods, Harefield; bog in Epping Forest; Hatfield forest; Barking; Hoddesdon Marsh; pond in Blackfan Wood, Bayford; Ruislip reservoir.

Carex divulsa.—Moist pastures and ditches, 5–6. About Brockham; Betchworth; Mickleham; between Ditton and Claygate; ditch by the roadside between Kingsbury and Hill Farm* (E. de C.).

Carex depauperata.—Dry woods, rare, 5–6. Charlton Wood; woods near Godalming; chalk-pit near Effingham (W. Reeves).

Carex distans.—Muddy salt-marshes near the sea, 6. Possibly on the banks of the Thames below Greenwich and Tilbury; "reports from inland stations erroneous, *C. binervis* mistaken for it" (H. C. Watson).

Carex elongata.—Marshes, rare. Meadows between the canal and river Weybridge, 6.

Carex flava.—Turfy bogs, frequent, 6. Plenty on Wimbledon Common and Hampstead Heath.

Carex fulva.—Boggy meadows, not common. Field near Harrow Weald Common; (doubtful species, Bot. Gazette) Little Berkhampstead.

Carex glauca.—Moist meadows and moors, common. Pinner;* Stanmore;* Mickleham Downs;* plenty, 6.

Carex hirta.—Wet pastures, woods, and ditches, frequent (Wimbledon Common;* Putney Heath*), 6.

Carex intermedia.—*See* C. disticha.

Carex lævigata.—Boggy thickets, rare, 6. In a thicket near Warley Common; Windsor Great Park; woods near Tunbridge Wells, frequent; wet places on Stanmore Heath; Spring Copse, Burgate, near Godalming; meadow among the Willows near Mortlake.*

Carex muricata.—Marshy and gravelly pastures, common (waste ground by the Paddington Canal*), 5–6.

Carex Œderi.—In similar situations. Bell Bar bog, Herts, not frequent, 5–6. Warley Common; Hertford Heath; bogs at Little Berkhampstead; Hatfield woodside and Kentish lane, Hatfield Park; by Pembridge lane; Hoddesdon new marsh; Holmwood; Whitemoor Common.

Carex ovalis.—Bogs and marshy places, common; (plenty about the bog on Hampstead Heath*), 6.

Carex pulicaris.—Bogs, frequent, 5–6. Putney Heath;* Epping Forest, near High Beech; Hainault Forest, near Fairlop; Shirley Common; Blackfan Wood, Bayford; bog at Little Berkhampstead; at Hatfield woodside; marshes, Stanborough; Hoddesdon Marsh; Bell Bar bog;

Warley Common; swampy copse near Hampton Lodge, Puttenham* (E. de C.).

CAREX PILULIFERA.—Moors, frequent, 6. Wimbledon Common;* Colney Heath;* Coulsdon Common;* Hayes Common;* Warley Common; Epping Forest.

CAREX PRÆCOX.—Dry pastures and heaths, not very frequent. Hayes Common;* Richmond Park;* Box Hill;* Warley Common; Epping and Hainault Forests; Ruislip.

CAREX PALLESCENS.—Marshy places, not very frequent, 6. Oxhey Wood;* wood by Pinner Lane, Watford (? the same); marshy meadows below Merstham pools; Warley Common; Epping Forest near Hale End; woods S. of Hertford, also in Broxbourne, Wormley, and Hoddesdon woods; by the lake, Bentley Priory; bogs above Peslik, Shiere; copse between Hook village and grounds of Ruxley Lodge, Claygate; Spring Copse, Burgate.

CAREX PANICEA.—Marshy places and bogs, common (Hampstead Heath;* Putney Heath;* Stanmore Marsh;* Harrow Weald Common),* 6.

CAREX PENDULA.—Ditches in damp cool woods, not frequent, but plentiful where it occurs, 5–6. Wormley Wood,* abundant; Alders Copse, Whetstone; Pinner Wood;* Ruislip;* Hatfield Forest; Warley Common; woods S. and S.W. of Hertford.

CAREX PSEUDO-CYPERUS.—Sides of ponds and ditches, not very general by the ditch on Barnes Common;* in a pond near Edgware;* Felbridge; boggy thicket, Warley Common; ponds S. and S.W. of Hertford, 6.

CAREX PALUDOSA.—Banks of rivers and ditches, common. (Ditch by the Thames between Hammersmith and Putney;* pond near Finchley).*

CAREX REMOTA.—Damp woods, swamps and ditches, in shady places, frequent (ditches about Kingsbury;* Esher;* Putney Heath*), 6.

CAREX RIPARIA.—Banks of rivers, common (banks of the Thames;* and Lea,* plenty), 6.

CAREX STELLULATA.—Bogs, common (Putney Heath;* Hampstead Heath*), 5–6.

CAREX STRICTA.—Bogs and sides of rivers, not common; unless mistaken for *Carex acuta*. A tuft or two on Putney Heath;* by the Roding;* Plaistow marshes.

CAREX STRIGOSA.—Shady woods and lanes, rare, 6. Moss Lane, Pinner; woods about Farnham; lane near Woodford; Lambourne Parsonage; Quicks Hill Wood, Hertford Heath; Blackfan Wood, Bayford, and Essendon Glebe woods; watercourse E. side of Brickendon lane; Hoddesdon, Wormley, and Broxbourne woods; Walton and Weybridge commons.

CAREX SYLVATICA.—Shady woods, frequent, 5–6. Harefield;* Coulsdon;* forests of Epping and Worley; Hatfield.

CAREX TERETIUSCULA.—Boggy and watery meadows, not common, 6. Wimbledon Common, west of the mill in one place;* by the side of the Paddington canal near Willesden Junction, a tuft or two;* near Epping; ponds in Broxbourne woods.

CAREX VULPINA.—Ditches, &c., common, and by the side of rivers; (plenty by the Thames,* and Lea, &c.*), 5–6.

CAREX VULGARIS.—Marshes and wet pastures, common (Hampstead Heath;* Putney Heath;* Ruislip moor*), 5–6.

CAREX VESICARIA.—Bogs and river sides, not common, 5–6. Pryor's Wood, Hatfield Heath, in a swamp* (E. de C.); Colney Heath; Reigate Heath? (*C. ampullacea* mistaken for it?) Ruislip reservoirs; near Hersham; Elstree reservoir; peat bog, Moor Park, Farnham; by the Roding; Barking; by the Mole (Sidlaw Bridge); in a pond below Bisham Wood; in Bayford Wood; road from Bayford to Little Berkhampstead, right side; in pools of the Colne near Colney Heath; between Gracious Pond and Woking station.

CARLINA VULGARIS. Dry hilly pastures in the chalk districts abundant, rare elsewhere, 6–9. Box Hill and the Betchworth hills;* hills W. of Dorking,* and in many other places along the range;* Banstead Downs;* Burnham Beeches; Rusthall Common; Tring; St. Albans; Great Berkhampstead Common.

CARPINUS BETULUS. Woods and hedges, also in plantations, common; (abundant in Epping Forest and on Warley Common*), 5.

Carum Carui.—Incidentally, near gardens and habitations. Abundant in Hyde Park near the Albert memorial* (E. de C.); railway bank, Pinner; waste ground, by canal, W. Drayton.

Castanea vulgaris.—Plantations, frequent about London in enclosures, 5–7.

CATABROSA AQUATICA. Pools and ditches, frequent (Barnes Common,* &c.; ditches by the Thames,* Lea,* &c.), 7–8.

CAUCALIS DAUCOIDES.—Cornfields in a chalky soil, not very common, 6. About the Betchworth hills; Banstead Downs; about Dartford, Erith, Northfleet, and Gravesend; roadside from Dorking to the Epsom Downs; Broomfield; summit of Buckland Hill in fallow fields; near Hitchin, plenty.

CENTAUREA NIGRA.—Meadows, pastures, and waysides; common everywhere (plenty by the banks of the Paddington canal*), 6–9.

CENTAUREA SCABIOSA.—Barren pastures, cornfields, and roadsides, especially in the chalk districts; plenty about Sutton,* and the Banstead Downs,* also in cornfields, Surrey side of the Thames opposite Teddington,* 7–9.

CENTAUREA CYANUS.—Cornfields, frequent, but not abundant. Cornfield near Foots Cray;* Caterham Junction;* Essex cornfields;* cornfield, summit of Reigate Hill;* about Hertford; Watford.

CENTAUREA CALCITRAPA.—Gravelly commons and waste places near the sea not frequent, but generally plentiful where it occurs, 7–8. Barnes Common, plenty;* near Northfleet by the roadside* (E. de C.); about Woodford (*olim*)?; West Ham; Plaistow;* Tilbury Fort; Dartford.

Centaurea solstitialis.—Rare, 7–9, in fields incidentally. Among lucerne at Essendon Glebe.

Centranthus ruber.—Old walls and ruins, and on chalk cliffs; plentiful in the Greenhithe and Northfleet chalk-pits;* old chalk-pits, Dartford; old walls between Foots Cray Church and Hurst (*olim*)? old chalk-pit by the Higham Station,* 6–9.

CENTUNCULUS MINIMUS.—Moist, sandy, and gravelly fields, not very common, perhaps often overlooked; Barnes; Shirley Common; between Frant and Tunbridge Wells by the road; low marshy ground by the paper mills, Hounslow; Chislehurst Common; High Beech, Epping Forest; about Hampton Court; Reigate Heath* (scarce); Farnham Common, where the turf has been cut; Gerard's Cross Common; Colney Heath; Bagshot Heath* (E. de C.); Esher Common, opposite front of Claremont; by a pond near the gravel fields between Hersham Green and St. George's Hill, 6–7.

CEPHALANTHERA ENSIFOLIA.—Beech woods on the chalk range, rare, local, 5–6. Between Mickleham and Headley in a wood left; borders of Cobham Park, or woods beyond, towards Cuxton?

CEPHALANTHERA GRANDIFLORA.—Beech woods on the chalk range; frequent, 5–6. Reigate Hill;* hills E. of Merstham * (near Caterham); hills E. of Shoreham;* about Gatton, Chipstead, Mickleham,* Box Hill; Hill Hall woods.

CERASTIUM ARVENSE.—Cornfields in a sandy soil, frequent; near Ham;* about Weybridge;* slopes of Chobham ridges;* about Chobham;* fields at the foot of the hills E. of Wrotham,* 4–8.

CERASTIUM GLOMERATUM.—Fields and roadsides everywhere, common,* 4–9.

CERASTIUM TRIVIALE.—Pastures, waste places, and wall-tops, common,* 4–9.

CERASTIUM SEMIDECANDRUM.—Waste places and wall-tops, not frequent, perhaps often overlooked. Shirley Common;* Hampstead Heath, 3–5.

CERASTIUM PUMILUM.—Rare, dry banks near Croydon (Curtis), subject of controversy; *vide* Cyb. Brit. i. 230, and app. iv.

CERASTIUM TETRANDRUM. Waste ground, walls and sandy places near the sea, local, 5–7. Tilbury; Southend; and Shoebury Common.

CERASTIUM AQUATICUM.—*See* STELLARIA AQUATICA.

CERATOPHYLLUM AQUATICUM.—Slow streams, ditches and pools; Thames near Hampton Court, river Lea about Tottenham and Chingford;* pond on Ditton Marsh;* and in Epping Forest; Hatfield Forest; many places between Hertford and Ware, in the Lea and adjoining ditches; ponds at Roxeth; Ruislip; Thames and adjoining drains about Walton Bridge and Sunbury Lock; Gatton Pond in the Park; in the Mole near Esher Mills, frequent (var. *submersum* is rare), 6–7.

CHÆROPHYLLUM ANTHRISCUS.—Waste places and roadsides, especially near habitations. (Plenty about Fulham towards Hammersmith,* and by the roadside from Barnes to Mortlake;* also near Barnes Common, on a bank right from Hammersmith.)*

CHÆROPHYLLUM SYLVESTRE.—Hedges and borders of fields, common; (plenty in hedges about West-end railway station * and lanes about Tottenham *), 4–6.

CHÆROPHYLLUM TEMULUM.—Hedges and roadsides, common (in lanes about Tottenham;* &c., with the preceding), 6–7.

Chærophyllum sativum.—A garden waif incidentally near habitations? 6.

Cheiranthus Cheiri.—Old walls and ruins about towns and villages; walls about St. Albans and Waltham Abbey,* 4.

CHELIDONIUM MAJUS.—Waste places about villages &c., frequent. 5–8. Moulsey,* Godalming,* &c.

CHENOPODIUM BONUS-HENRICUS.—Waysides and waste places, frequent (Hatfield Forest; West Ham; Barking), 5–8.

CHENOPODIUM ALBUM.—Fields and waste places, common (everywhere in the London suburbs), 7–9.

CHENOPODIUM GLAUCUM.—Waste places, not very common; about London; in a waste place near West-end railway station;* Chingford Hatch, 8–9.

CHENOPODIUM OLIDUM.—Waste places and by roadsides; not common, 8–9. About Putney; Woolwich; Walthamstow; by the railway bridge near the station, Willesden;* Wimbledon Common road to Kingston in the hollow;* Purfleet; Southend; about St. Albans.

CHENOPODIUM RUBRUM.—Dunghills and under walls, 8–9; also in salt-marshes. Warley Common; Purfleet; Southend; Weston Green; Milwood Green; Hoddesdon old mill.

CHENOPODIUM HYBRIDUM.—Waste places and cultivated fields about London, 8.

CHENOPODIUM URBICUM.—Waste places near houses: 8. Outskirts of Epping Forest, between Hoddesdon and Hertford Heath; near Chobham, a var.

CHENOPODIUM FICIFOLIUM.—Dunghills and waste ground about London and Tilbury, 8–9.

CHENOPODIUM POLYSPERMUM.—Waste places and fields, on rubbish.

CHENOPODIUM MURALE.—Waste places near habitations. Hoddesdon, N. of the town towards Ware, 8–9.[1]

CHLORA PERFOLIATA.—Pastures and downs in the chalk districts. Croydon;* Box Hill and Betchworth hills;* Reigate Hill;* hills E. of Merstham;* hills W. of Dorking;* E. of Shoreham;* E. of Wrotham,* abundant; stone chalk pits, Greenhithe, do.;* Tring downs;* Harefield; Colsdon; Hog's Back;* chalk-pits E. of Hatfield Park; Stanstead, borders of Thrift Wood.

CHRYSANTHEMUM LEUCANTHEMUM.—Dry pastures and railway banks, common everywhere (in profusion about Wandsworth; Clapham Junction &c.; Paddington Canal banks *), 7–9.

CHRYSANTHEMUM SEGETUM.—Cornfields, frequent; especially in Herts and in Essex near Wormley west end, in profusion;* field leading to Telegraph Hill also in abundance;* near Colney Heath, towards Springfield station,* 6–9.

CHRYSANTHEMUM PARTHENIUM.—*See* MATRICARIA.

CHRYSANTHEMUM INODORUM.—*See* MATRICARIA.

CHRYSANTHEMUM CHAMOMILLA.—*See* MATRICARIA.

CHRYSANTHEMUM TANACETUM.—*See* TANACETUM.

CHRYSOSPLENIUM ALTERNIFOLIUM.—Boggy places near springs, rare, 4–6, Littleton bridge, Reigate; and in lane by Wonham Mill; in an alder swamp S.E. of Reigate Heath, &c.;* water-courses near Unsted Bridge; Epping Forest (near Epping); Thrift Wood, Chelmsford; Clifden Wood near Taplow.

CHRYSOSPLENIUM OPPOSITIFOLIUM.—Sides of rivulets in woods and in similar situations as the preceding, rare; Epping Forest; meadow below

[1] Species difficult to determine; any number may be made. *Vide* Moquin, in vol. iii. of De Candolle's 'Prodromus.'

Coney Farm, Harefield; moist copses between Shelford and St. Martha's Chapel, Guildford; Wonham, near Buckland; in an alder copse with the preceding near Reigate Heath;* swamps below the hills E. of Merstham; by brook, S. end of Broxbourne Wood; Wormley Wood; banks of brook near Theobalds, Cheshunt; wood near Beaumont Green; Quick's Hill Wood, Hertford Heath; plentifully by a spring in Bramfield Brook bottom; near Totteridge in a spring near Bourne End, Watford; near Claygate; near Walton Bridge; swamp near Redhill railway station, 4–6 Bishop's Wood, Highgate; foot of Boar Hill.

CICENDIA FILIFORMIS.—Rare. Tilgate Forest, bog between Pease Pottage gate and Starvemouse Plain.

CICHORIUM INTYBUS.—Borders of fields, waste places, and roadsides especially in the chalk districts, frequent, 7–10. (Plenty between Carshalton and Banstead.*)

CICUTA VIROSA.—Swamps and pits, rare. Swamp in Wormley Wood on its western border;* swampy pool near Appleby, two miles W. of Cheshunt * (E. de C., 1876); (localities in Middlesex erroneous? 'Cybele,' *Œnanthe fluviatilis* mistaken for it?) ponds near Brickendon Green, Herts? (No longer there.)

CINERARIA.—*See* SENECIO CAMPESTRIS.

CIRCÆA LUTETIANA.—Shady woods and lanes, common; White Hart Lane, Tottenham;* Telegraph Hill, Ditton, and lane leading from Esher to the lower part of Winter Downs,* 6–8. Epping Forest, near Woodford.*

Claytonia perfoliata.—An alien; incidentally. Turf-bank near the windmill, Wimbledon; hedgebank, Weybridge Common.

CLEMATIS VITALBA.—Hedges and copses, chiefly in the chalk districts; abundant, 7–9. Plenty about Croydon, in roads leading to Selsdon and Sanderstead; Riddlesdown.

COCHLEARIA OFFICINALIS.—Rare, local, by the seashore, Southend,* in no great quantity to the W. of it, 5–8.

COCHLEARIA ANGLICA.—Muddy shores of the Thames below Woolwich, on both sides, in several places;* between Greenwich and Woolwich.*

COLCHICUM AUTUMNALE.—Meadows and pastures, rare, 8–10. Meadow left of the London road adjoining W. end of Wray Common.

COMARUM PALUSTRE.—Marshes and bogs, local, rare, in S.W. Surrey only, 5–7. Reigate Heath? not seen there; Keston Common? (*olim*); ditches in Pound Lane, Epsom? (not now); Wormley Wood, in the swamp with *Cicuta?* (*olim*); Rickmansworth Common Moor? In bogs on the common near Tilford * (yes, and plenty on the borders of a pond there;* above what was formerly Abbots Pond, now a pasture); N.E. corner of a pond on Puttenham Common, near some alders;* Ball's Wood, Herts, pond N. end; Pryor's Wood, Hertford Heath?

CONIUM MACULATUM.—Waste places, ditches and hedge borders in damp shady localities; frequent; 6–7. Old lane leading from Kingsbury to Whitechurch;* hedge in the hollow between Neasdon and the railway bridge right, in a blind lane;* Ruislip Moor;* hedge-side in a field close to Park Station, Tottenham;* swamp near Merstham; Boxhill;* Shiere; Albury.

CONVALLARIA MAJALIS.—Woods and coppices, in cool damp places, 5–6.

Croham Hurst, near Croydon;* about Coulsdon; Guildford; copse near Worplesdon; wood near Chislehurst, left before reaching the place; Darent Wood;* wood on Winchmore Hill; Farnham Common, under Cæsar's camp; Ongar; Epping Forest, near High Beech; near Warley Common; Hurtwood Common; Reading (in an island); Clifden Wood; Pryor's Wood, Hertford Heath; wood near Leggatts, Northaw; Wormley Wood (on Foulwell's Farm); Stubbins' Wood, S.W. of Tring.

CONVALLARIA MULTIFLORA.—*See* POLYGONATUM.

CONVOLVULUS ARVENSIS.—Cornfields and roadside hedges, common, everywhere, 6–7.

CONVOLVULUS SEPIUM.—Moist woods and hedges, common. (Lanes about Tottenham and Edmonton, plenty.)

CONVOLVULUS SOLDANELLA.—Local, seashores; on the shore at Southend.

Coriandrum sativum.—Occasionally by the Thames, rare, elsewhere incidentally; lane between Dorking and Ranmore; Southend; Thames bank between Greenwich and Woolwich; a plant or two.*

CORNUS SANGUINEA.—Hedges and thickets, especially in the chalk districts, also in plantations; (plentiful about Dorking; Box Hill and along the chalk range, Croydon towards Selsdon &c.), 6–7.

CORONOPUS.—*See* SENEBIERA.

CORYDALIS CLAVICULATA.—Bushy and shady places on or near a gravelly soil, 6. Shooter's Hill in a gravel pit? (*olim*); about Coulsdon; Grays Wood, near Godalming; wet copses about Reigate Heath;* moist woods about Abinger and foot of Leith Hill; about the north base of Boar Hill; sandy ground near the old camp, Wimbledon* (hedgebank); Keston Common; High Rocks, Tunbridge Wells; about Shiere and Albury.

CORYDALIS LUTEA.—Old walls, not unfrequent, Highgate;* Greenwich;* Eltham;* St. Albans Abbey (*olim*).

CORYLUS AVELLANA.—Woods, copses and hedgerows, common everywhere, 2–4.

COTYLEDON UMBILICUS.—Walls and housetops, incidentally, 6–8, not frequent. Devil's Jumps, (on cottage walls) near Frensham.

CRATÆGUS OXYACANTHA.—Woods and hedges, common everywhere, 5–6. Var. *oxyacanthoides*, Hatfield.

CREPIS BIENNIS.—Chalky pastures, rare. Boxhill; Banstead Downs, near Ewell; between Gravesend and Rochester; Morant's Court Hill.

CREPIS FŒTIDA.—Dry chalky ground, 6–7, rare. Banstead Downs (?); about Greenhithe; Northfleet; Purfleet; Gravesend; Grays; clover field, Chessington; here and there on the chalk from Knockholt to Wrotham; old quarries W. of Dorking; field behind Juniper Hill, Mickleham.

CREPIS TARAXACIFOLIA.—Chalky pastures on the North Downs not unfrequent; Sutton near the station (Topogr. Bot.); Cobham, Surrey; Greenhithe;* chalky banks, foot of the downs E. of Wrotham * (E. de C.); Purfleet; Grays; Leigh; Erith; Dorking chalk-pits.

CREPIS PALUDOSA.—Moist woods and rocky places (?). Marshy meadows below Woolwich;* and near Erith,* plenty; local.

CREPIS VIRENS.—Dry pastures; walls; roofs; hedges; everywhere, common, 6–9.

Crocus vernus.—Meadows and fields, rare. Meadows at Totteridge; Brookman's Park, N. Mimms, 4–5.

CUSCUTA EPITHYMUM.—Parasitic on *Calluna*. Heaths, frequent; on all the Surrey heaths;* Wimbledon Common.*

CUSCUTA EUROPÆA.—Parasitic on nettles, thistles, &c., rare, 7–8. On nettles, foot of Box Hill? (*olim*); Brockham (?) by the Mole (*olim*); Epping Forest; Warley Common; osier holt by the road a little above Guildford; at Reading, on lucerne; near Chertsey and at Maidenhead, abundant (*olim*); by the Mole, Cobham.

CUSCUTA EPILINUM.—On flax, rare, 8. 'Cybele Brit.' ii. iii.

CUSCUTA TRIFOLII.—On clover, rare, 7–9. Fyfield; Winter Hill, Berks; top of Reigate Hill; Redhill; and High Trees Farm, Brockham; Guildford.

CYNOGLOSSUM OFFICINALE.—Waste ground, and by waysides, especially in the chalk districts, 6–7. Mickleham Wood, above the church;* about Dartford;* (in the marshes) Greenhithe and Northfleet; Erith;* Gravesend;* plenty, banks of the canal, a mile or two towards Higham (E. de C.); Southend; Smitham Bottom (?); Guildford; Farnborough; Keston Common; Purfleet; Richmond Park; Trumpets Hill, Reigate; Hatfield Park; Ware Park; near Ware, by the London road.

CYNOGLOSSUM MONTANUM.—Shady woods and roadsides, not common, more frequent in Essex, 6–7. Between Chingford and Walthamstow (*olim*); Larks Wood, Chingford, and elsewhere perhaps in Epping Forest; about Fyfield;* Hatfield Forest* (E. de C.); Hainault Forest (or in copses the remainder of it); Southend, near Eltham (*olim*); Braxted, by the roadside; below Whitehill, hills E. of Merstham* (E. de C.); in Norbury Park, plenty; Purfleet, in the wood there; Cashiobury Park.

CYNOSURUS CRISTATUS.—Pastures and roadsides, common, 7. (Plenty about Hendon.*)

Cyperus fuscus.—Rare, local, 8–9. Borders of one of the ponds on Shalford Common; plenty.*

DACTYLIS GLOMERATA.—Pastures and roadsides, common everywhere, 7–8.

DAPHNE LAUREOLA.—Woods and thickets, frequent in the chalk districts, 2–5. Mickleham; Box Hill and the Betchworth hills; Epping Forest; woods about Harefield; on the Hog's Back;* Long Valley Wood, Rickmansworth; woods E. of Shoreham;* in a grove at Breakspeares, Harefield.

DAPHNE MEZEREUM.—Woods, rare; about Coulsdon; Boxhill; near Stroud House, Godalming; Bisham Wood, Berks, 3–4.

DAUCUS CAROTA.—Pastures, borders of fields, and roadsides, common, everywhere; especially in the chalk districts, 6–7. (Plenty between Carshalton and Banstead.)

Delphinium Ajacis.—Sandy or chalky fields, and incidentally near gardens and habitations, 6–7. Cornfields in Ditton parish; cornfields about Reigate (*olim*); and about Croydon (*olim*). Cornfields, Hoddesdon; by the stream at High Rocks.

DENTARIA BULBIFERA.—Shady and moist woods, rare; Old Park Wood Harefield,* abundant (enclosed), also in Garret Wood, Pinner side of

Harefield; near Croydon; in a wood near Wallington (*olim*); on rocks by the rivulet at High Rocks, Tunbridge Wells, plentiful, and elsewhere in the neighbourhood; wood adjoining High Wood at Rickmansworth, 4–5.

DIANTHUS ARMERIA.—Pastures and hedges, not frequent. About Coulsdon; road from Mickleham to Dorking? (*olim*); Charlton Wood, *olim*; about Dartford; Darent Wood; between Shorne and Stroud; about Eltham; Bromley; Woodford; Harefield; Albury; between Cobham and Cuxton; bushy places on the high bank between Southend and Leigh; lane by Trumpets Hill, Reigate; and on Redstone Hill; Epping Forest; Frith Hill, Godalming; about Hertford and Hatfield; gravel pit, left of the road from Howell Farm to Hatfield; and abundantly near old pit short of this; gravel pit, Hertingfordbury road; Mangrove Lane; wood between Panshanger House and Hertford Lodge; bridle way from Great Berkhampstead to Bovingdon, near Bottom Farm; Denbies Hill, Dorking; footpath between Guildford and Albury.

DIANTHUS DELTOIDES.—Gravelly pastures, rare, Thames side near Teddington Lock, sparingly;* about Coulsdon; Duppas Hill, Croydon; Hampton Court; Totteridge Green, back of Osmund's Barn, 6–9.

DIANTHUS PROLIFER.—Gravelly pastures, rare, between Teddington and Hampton Court? (*olim*), 6–10.

DIGITALIS PURPUREA.—Dry banks and woods in hilly situations (not frequent in the chalk districts except in deep overlying gravel drift), common. Borders of Coombe Wood, Wimbledon;* Croham Hurst; Croydon;* Chislehurst; Keston Common; Epping Forest;* Warley Common;* Darent Wood;* woods on the flanks of Leith Hill;* lanes W. of Dorking;* Harrow Weald Common.

DIGITARIA HUMIFUSA.—Sandy fields, rare, about Weybridge? (*olim*); about St. Martha's Chapel, Guildford, 7–8.

DIGRAPHIS ARUNDINACEA.—Sides of ponds, rivers, ditches, common everywhere. (Plenty on the banks of the Lea and canal;* Thames and ditches near.*)

DIPLOTAXIS TENUIFOLIA.—Waste places by the Thames, and in chalk-pits, abundant, less frequent elsewhere; plenty in the Greenhithe and Northfleet chalk-pits;* by the Thames in several places below Woolwich on both sides;* also above London, occasionally; Moulsey;* Sunbury,* 6–9.

DIPLOTAXIS MURALIS.—Sandy fields, rare, Tilbury; Southend; Watford; 8–9.

DIPSACUS SYLVESTRIS.—Hedges, roadsides and waste places, common, 7 (frequent about London; railway banks near the Finchley Road;* by the Thames at Putney),* 8–9.

DIPSACUS PILOSUS.—Moist hedges, rare. Hedges between Wanstead and Barking? (*olim*); wood between Chislehurst and Orpington; about Guildford plentiful; near Moor Hall, Harefield; Beeching Wood, Norbury Park; Ball's Wood, Herts, and ditches near* (E. de C.); in a shady lane leading from Wrotham to Cuxton, about 1½ miles E. of Wrotham* (E. de C.); between Chingford and Waltham Abbey; near Chilworth; Sonning Lane; Roydon Lane, Stanstead; Bayford Wood; Essendonbury

Lane, S. of Brickendonbury, &c.; in the wood near the boggy meadow, back of high grounds, Hoddesdon.

Doronicum Pardalianches.—Incidentally in damp and hilly woods, rare, 5–7.

Doronicum plantagineum.—Incidentally, in damp places, rare, 6–7. Fyfield; Shooter's Hill, left of lane through West Wood and among the trees? (1848); wood near Chislehurst? (1848).

DRABA VERNA.—Walls and dry banks, common (on old mud-topped walls, everywhere in the suburbs of London),* 3–6.

DROSERA INTERMEDIA.—Bogs, mostly on peaty heaths and chiefly in Surrey; Esher Common;* Chobham Common;* Pirbright Heath;* Bagshot Heath;* Elstead and Puttenham Commons;* bogs about Farnham; near Felbridge; Ruislip and Harrow heaths; Warley Common? Harefield? (*olim*); Burnham Beeches, bogs near, 7–8.

DROSERA ROTUNDIFOLIA.—Bogs generally, and on moist heaths. Common in Surrey; scarce in Herts, Essex and Middlesex. Plenty in the bog on Hampstead Heath;* Putney Heath;* Esher Common;* Reigate Heath,* &c.

Echinochloa Crus-galli.—Fields near London? (*olim*) rare, 7.

ECHIUM VULGARE.—Banks, fields and waste ground, especially in a sandy or gravelly soil, very common in the chalk districts, rare elsewhere. Everywhere on the Surrey and Kentish downs,* 6–7. Plentiful between Carshalton and Banstead.*

ELATINE HEXANDRA.—Margins of ponds, rare, local, 7–9. Lowermost of the Cutmill ponds on Puttenham Common; Frensham Pond; Silk Mill Pond on Thursley Common; Dam Head Cascade, Virginia Water; Felbridge Pools; Hedge Court millpond, S. side; small pond on the heath not far from Pirbright.

ELATINE HYDROPIPER.—Ponds with the other on Puttenham and Thursley commons, rare, local, 7–9.

Elodea canadensis.—Ponds and slow streams, common, everywhere. A doubtful alien (Cybele Brit.). Plentiful in the Lea Canal;* also in the Pond on Hampstead Heath.*

ELYMUS ARENARIUS.—Sandy seashores, rare, local; on the shore below Southend; towards Shoeburyness, 7.

EPILOBIUM ANGUSTIFOLIUM.—Moist banks, and margins of damp shady woods, 7. Not frequent, but plentiful where it occurs. Summit of Box Hill;* railway banks beyond Weybridge;* Broxbourne and Wormley woods* (E. de C.); railway cutting, Brentwood; Holmwood;* Great Berkhampstead and Frithden copse, Berkhampstead Common.

EPILOBIUM HIRSUTUM.—Banks of rivers, ditches and ponds, common, 7–8. (Abundant by the Lea.*)

EPILOBIUM PARVIFLORUM.—In similar situations with the above, less frequent; marshy places by the Basingstoke canal, 7–8.

EPILOBIUM MONTANUM.—Shady banks and roofs, also on walls, common. (Frequent on the Harrow and Edgware roads, and about Willesden, &c.*)

EPILOBIUM TETRAGONUM.—Sides of ditches, and watery places, common. Plenty in the ditches by the Lea and Canal.*

EPILOBIUM ROSEUM.—Sides of rivers and edges of milldams, &c., not common, 7–8. Moss Lane, Pinner; about Abrook; about Shiere; between Moreton and Ongar; near Farnham by Bowen mill, in a wet lane leading to Aldershot; Rickmansworth; bank of a ditch by the foot-path to Scott's Bridge; near Totteridge (? by the Brent); in the lane from Hatfield town to the Union Workhouse; between Waltham Abbey and Epping; Brentwood; Wimbledon Common; near Albury Church, by the river; Bayford, near the Church, in a ditch by the Hertford road; brook near Theobalds, Cheshunt; between Gracious Pond and Woking station.

EPILOBIUM OBSCURUM.—In similar situations with *E. tetragonum*, of which till lately it was considered a variety. Hatfield; Harefield; Harrow Weald Common; Stanmore Heath; Mill Hill; Colney Heath.

EPILOBIUM PALUSTRE.—Boggy places and by the sides of ponds and ditches, frequent, 7–8. Putney Heath;* plenty by the Canal near Woking station.*

EPIPACTIS LATIFOLIA.—Woods in hilly countries, generally, if not solely in beech woods on the chalk; not uncommon, 7–8. Box Hill;* Mickleham woods;* Reigate Hill;* hills E. of Merstham, towards Caterham;* Gatton; Chipstead and Buckland hills; Highwood; Rickmansworth; woods S.W. of Tring; Albury Nowers Wood;* woods on the chalk about Farnham Woods, Warley; about Shiere; Bisham, Herts; woods at Essendon; Broxbourne Wood; Undermole Wood, Hertford; Bedwell Park and plantations, thence to Little Berkhampstead; Harrow Weald Common.

EPIPACTIS PALUSTRIS.—Marshy places, mostly in a chalky soil or subsoil, rare, 7. Below the Betchworth hills; near Hemel Hempstead, coming from Watford, plentiful; Wormley Wood, in a boggy pasture;* S. side Merstham pools (among the willows); Wargrave Hill; wet pasture by the brook near East-end Green, Hertingfordbury.

EPIPACTIS PURPURATA.—An ambiguity. Reigate; King's Wood Farm; in shaws; weald clay below Crawley.

Eranthis hyemalis.—Incidentally; rare, 4. Albury Park.

ERICA CILIARIS.—Heaths, rare. Frensham or Farnham.

ERICA CINEREA.—Heaths and commons, frequent, 7–9. Abundant on all the Surrey heaths;* rare in Herts, Essex and Mids.; Hampstead Heath, sandy parts;* Abrook Common, towards Oxshott Hill.*

ERICA TETRALIX.—Moory and boggy ground on heaths, frequent. Surrey heaths all;* Hampstead Heath; Esher Common;* Putney Heath;* rare in Herts and Essex and Mids., 7–8.

ERIGERON ACRIS.—Dry gravelly pastures, chiefly in the chalk districts, frequent, but nowhere in profusion (E. de C.), 7–8. Chalky banks near Croydon; Leatherhead; Dartford; Farnborough; Erith;* near Cobham;* Greenhithe; Shoreham near the station;* and chalky banks below hills E. of Shoreham;* W. of Dorking;* Reigate Hill;* Croham Hurst; and chalk-pit (borders of) near;* chalky banks, hills W. of Dorking, sparingly;* Purfleet; between Keston Common and Down; Box Hill; lower slopes of Buckland Hill; Sonning; footpath between Hertford and Bayford; also between Bayford Hall and Bayford Wood; gravel pit near Hertford by the Welwyn road; between Hersham and St. George's Hill.

Erigeron canadensis.—Gravelly wastes, frequent 8–9, about London and the London Railway banks and stations in many places,* also near Erith;* Walthamstow,* by the reservoirs; also on cultivated ground in the suburbs.*

ERIOPHORON ANGUSTIFOLIUM.—Turfy bogs and moors, common. Bog on Hampstead Heath;* Sunning Hill bog near Ascot, in profusion.*

ERIOPHORON VAGINATUM.—Turfy bogs and moors, by itself or with the above, not so general. Sunning Hill bog near Ascot, in profusion;* Reigate Heath;* Leith Hill;* Woking Heath;* and bogs by the Canal, Brookwood;* Chobham Common;* Witley Common,* 3–5.

ERIOPHORON GRACILE.—Bogs, rare. Whitemoor Pond, between Woking and Guildford? (one of the two ponds on the common at Whitemoor [Worplesdon?] has been drained and converted into a pasturage; no Eriophoron by the other), 6–7.

ERYTHRÆA CENTAURIUM.—Dry pastures, frequent, 6–9. Everywhere on the Surrey and Kentish downs;* in great abundance and luxuriance in Wormley Wood;* Epping Forest; Keston Common;* Northfleet; Tring;* Ball's and Bayford woods.

ERYTHRÆA PULCHELLA.—Rare. Between Hertford and Bayford; chalk-pits, Purfleet; in a meadow since converted into a wood, and also in the second field from East-end green, Hertingfordbury, by the footpath to Watery Hall; cornfields between Rusthall Common and the road to High Rocks, Tunbridge Wells; banks of ponds near Three Bridges station.

ERYNGIUM MARITIMUM.—Sandy sea-shores near Southend;* plenty, 7–8.

ERODIUM CICUTARIUM.—Waste ground, frequent, 6–9. Shirley Common;* beyond Twickenham;* Barnes Common, bank near the station.*

ERODIUM MARITIMUM.—Sandy sea-coasts, rarely elsewhere (sandy bank one mile from Farnham! Bot. Gazette), 5–9.

ERYSIMUM CHEIRANTHOIDES.—Waste places, and fallow as well as cultivated fields, frequent. About London in gardens;* railway banks and stations,* Clapham Junction (*c. g.*); banks of the Thames, opposite Sunbury;* field between Weybridge and Chobham Common, left, in abundance;* Wandsworth.*

ERYSIMUM ALLIARIA.—*See* SISYMBRIUM.

EUONYMUS EUROPÆUS.—Woods and thickets, frequent; but nowhere in abundance, 5–6; more general on the chalk range. One tree in a lane leading from Hampstead Heath to Fortune Green;* Purfleet; Harrow Grove, plenty; in bushy parts of Wimbledon Common;* lane leading from Willesden to Acton;* Epping Forest;* woods between Stoke Common and Burnham Beeches;* wood south of Chislehurst Common, east,* &c.; woods about Godalming and Reigate.

EUPATORIUM CANNABINUM.—In watery places and river banks; frequent, 7–9. Banks of the Roding near Chigwell;* plentiful in the Colne and bordering ditches, between Rickmansworth and Uxbridge;* in the Wey;* Merstham pools; &c.

EUPHORBIA PEPLUS.—Waste and cultivated ground; common, 6–10. (Frequent about London; plenty in the market-gardens by the roadside between Hammersmith and Barnes Common.*)

EUPHORBIA AMYGDALOIDES.—Woods and thickets; common, 3–5. Frequent in Epping Forest;* in the patch between Woodford and Walthamstow;* Croham Hurst;* Darent Wood;* abundant.

EUPHORBIA EXIGUA.—Cornfields, common. Plentiful in cornfields about Sutton, Banstead, and Epsom, &c.*

EUPHORBIA PLATYPHYLLA.—Cornfields on the slopes of the Surrey and Kentish downs; not very frequent, 6–10. Near Northfleet; cornfields under hills W. of Reigate;* Fyfield; cornfields between Harefield and Bartleswell; about Farnham.

EUPHORBIA PORTLANDICA.—Local. (?) Rare, bogs in Charlton Wood (*olim?*); 5–9.

EUPHORBIA HELIOSCOPIA.—Cultivated and waste ground; frequent, 6–9. Plenty in some waste land, right of the road from Putney Heath to Kingston;* and in similar localities about London.*

Euphorbia Lathyris.—Thickets and underwoods, very rare. Woods near Cobham, Kent; Brentwood.

EUPHORBIA PARALIAS.—Rare, local. S. Shoebury Common.

EUPHORBIA STRICTA.—Local, rare; cornfields and among the tares. Farnham (Bot. Gazette).

EUPHRASIA OFFICINALIS.—Heaths, moory pastures, and roadsides in heathy districts; common, 7–9. Wimbledon Common (sides of drains and moory parts);* Surrey and Kentish downs and heaths, everywhere;* commons in Bucks, &c., abundant.*

FAGUS SYLVATICA.—Woods, especially on a chalky soil and in plantations; common, 4–5. Borders of Hampstead Heath, planted;* environs of London, planted;* abundant on the Surrey, &c., hills.*

FESTUCA BROMOIDES.—*See* F. SCIUROIDES and F. PSEUDO-MYURUS.

FESTUCA SCIUROIDES.—Dry pastures and on walls; frequent, 6. Blackheath; E. Tilbury;* Brentwood; Redhill; about Hertford; Brentford; Walton Bridge; Reigate; Purfleet;* Erith.*

FESTUCA PSEUDO-MYURUS.—In similar situations, but of less frequent occurrence, 6. Dry parts of Epping Forest; E. Tilbury; Purfleet.

FESTUCA UNIGLUMIS.—Rare; local; Southend.

FESTUCA OVINA.—Dry upland pastures, common, 6–7. (Hampstead Heath, abundant.)

FESTUCA RUBRA.—In similar situations less frequent. (A variety of the above; see 'Phytologist,' iii. p. 261.)

FESTUCA DURIUSCULA.—*See* F. RUBRA, of which it is a variety, and both of *F. ovina* (Hooker and Arnott).

FESTUCA ELATIOR.—Shady woods and river banks; frequent, 6–7. Plentiful by the banks of the Thames below Greenwich;* var. *arundinacea*, near Putney.*

FESTUCA PRATENSIS.—Moist meadows and pastures; common, 6–7. Plentiful by the Lea, &c.*

Filago gallica.—Incidental, rare. In a cornfield between Chilworth Wood and St. Martha's Hill; between Hertford and Welwyn.

FILAGO GERMANICA.—Sandy and gravelly places, and dry pastures; frequent, 7–9. (Putney Heath.*)

FILAGO APICULATA. In similar situations, of less frequent occurrence. Sandy places about Cobham and Esher; Wargrave in Berks; gravelly field on the Hertford road near N.E. boundary of Hatfield Park; near Weybridge station; fields about Fairmile; sandy field left between Hersham and St. George's Hill; sandy fields by roadside between Chertsey and Virginia Water; near Witley station.

FILAGO SPATHULATA.—In similar situations, rare. Road between Staines and Hampton Court; by the Thames near Twickenham church; several localities about Esher and Ditton.

FILAGO MINIMA.—Gravelly places; frequent, 6–9. Plenty further end of Wimbledon Common;* also on Wandsworth Common;* and on Putney Heath.*

FŒNICULUM VULGARE.—Chalky banks near the sea, also on railway banks incidentally, and waste places near habitations; local; Charlton chalk-pit (*olim* ?); about Dartford; the Crays (?) Northfleet;* and Gravesend (incidentally in several places);* Purfleet; Southend.

FRAGARIA VESCA.—Woods and hedgebanks in woodland districts; common, 5–6. (Plenty in Epping Forest.*)

Fragaria elatior.—Incidentally near habitations for the most part. Tilgate Forest.

FRANKENIA LÆVIS.—Sea-shores; local; rare; near Southend; Isle of Sheppey, abundant, 7–8.

FRAXINUS EXCELSIOR.—Woods and hedges; everywhere frequent, 4–5. Some trees on the Edgware Road and neighbouring lanes towards the Welsh Harp.*

FRITILLARIA MELEAGRIS.—Meadows and pastures in the Thames valley and its tributaries; local, not general, 4–5. Between Barnet and Hatfield; in a pasture near 15th milestone (*olim* ?); pasture near railway arches, Watford; moor meadows near Ruislip; in a meadow at Northaw; fields about Stroud, Godalming; about Pinner (abundant in one place; field right of road from the town towards Rickmansworth);* Totteridge; Hoddesdon marsh; near Mortlake and Coombe Wood (*olim*); Maidenhead; Reading.

FUMARIA OFFICINALIS.—Dry fields and roadsides; common, 5–9. Plentiful between Carshalton and the Banstead Downs.*

FUMARIA VAILLANTII.—Rare, local. Kent.

FUMARIA CAPREOLATA.—*See* F. PALLIDIFLORA. (Few records except under the aggregate name.)

FUMARIA PARVIFLORA.—Cornfields in a chalky soil; rare, 6–9. About Box Hill; between Dartford and Darent Wood;* Coulsdon; Epsom; Hog's Back;* Cobham.

FUMARIA MURALIS.—Rare (?) Hedgebank, Barnes, Surrey. (Report Bot. Ex. Club, 1876.)

FUMARIA PALLIDIFLORA ?—Cornfields, gardens, hedges, &c., very unfrequent, 5–9. [Recorded as *F. capreolata*; perhaps var. *Borœi*, or other segregate.] Field near Easney Park Wood, footpath to Ware (?); between Brockham and the downs.

FUMARIA MICRANTHA.—*See* F. DENSIFLORA.

FUMARIA DENSIFLORA.—Rare, 6–9. Cornfields on the Hog's Back.

GAGEA LUTEA.—Woods and pastures, rare; meadow near Godalming.

GALANTHUS NIVALIS.—Woods, orchards, and meadows; unfrequent, 3–4. Box Hill, *olim* (?); near Stoke Park, Guildford; meadows. Bourne-end Mill, near Watford; and in Lea meadows near Bracket Hall; on a farm 1½ mile beyond Banstead Park; Croham Hurst, near Croydon, *olim* (?); in a wood both sides of the way over Farden downs, near Chaldon; Chipping Ongar; Reading; meadows, banks of the Mole from Betchworth to Brockham; and both sides of a stream in a field right of a lane running S.W. from Brockham Green.

GALEOPSIS TETRAHIT.—Cornfields, hedges, and in cultivated ground; frequent but not in great abundance. Occasionally about London, as near Willesden;* banks of the Paddington canal;* between Radlett and Colney Heath;* and cornfields, Roxeth.

GALEOPSIS LADANUM.—Cornfields on the chalk range, plentiful; rare elsewhere. Cornfields on Reigate Hill;* Box Hill;* hills W. of Dorking;* W. of Caterham Junction;* about Riddlesdown; and about the Banstead Downs;* also cornfields in Essex, but less frequent; Hatfield, Broad Oak;* The Lavers;* The Rodings;* on the weald clay about Capel; Hoddesdon.

GALEOPSIS VERSICOLOR.—A boreal form of *G. Tetrahit* (Mr. Watson). Records, any (?) of its occurrence near London.

GALEOPSIS OCHROLEUCA.—Rare. Dorking chalk-pit; between Dartford Heath and Greenstreet Green; chalk-pit, Darent Wood.

Galinsoga parviflora.—Rare. Naturalised near Kew, E. Sheen, and Richmond.

GALIUM CRUCIATUM.—Hedgebanks and thickets; frequent, 4–6. On the borders of Darent Wood;* and on Keston Common;* copses and in hedges bordering Epping Forest;* &c.

GALIUM VERUM.—Dry banks and pastures; common, 6–9. Barnes Common;* about Croydon;* and between Carshalton and Banstead,* &c.

GALIUM ULIGINOSUM.—Wet meadows and sides of ditches; frequent, 7–8. (On Hampstead * and Putney heaths.*)

GALIUM SAXATILE.—Heathy and upland pastures; frequent, 6–8. (Plenty on Hampstead Heath.*)

GALIUM PALUSTRE.—Sides of ditches and ponds; frequent, 7–8. Putney (Heath and Wimbledon Common*).

GALIUM MOLLUGO.—Hedges and thickets; common, 7–8. (Wimbledon Common.*)

GALIUM APARINE.—Hedges; common, everywhere, 6–7.

GALIUM ANGLICUM.—Rare; old walls, &c., in the chalk districts, 7. Dartford; Croydon (Duppas Hill); Northfleet and Gravesend; Purfleet* (E. de C.); walls of Bayham Abbey (Tunbridge Wells).

GALIUM TRICORNE.—Cornfields in the chalk districts; not common, 6–8.

Below Buckland hills;* Croydon; Gravesend towards Cobham* (E. de C.); Hatfield, Broad Oak; Fyfield; Hoddesdon; fields about Hertford.

GALIUM ERECTUM.—Rare; Surrey chalk hills. Sandy field, Fairmile near Esher (H. C. Watson).

GASTRIDIUM LENDIGERUM.—Flats near the sea; about the margins of woods in Kent and Surrey (Cybele Brit.); and in woods half-way between Chelmsford and Braintree; rare, 6–10. Woolwich warren (?) *olim;* Gravesend; Rochester; Erith; Isle of Sheppey; margin of a wood on Telegraph Hill, Surrey;* field near the waste midway between Hatfield and St. Albans; about Thorndon; wheat-field near Oxshott Hill in 1867; not unfrequent in the Wealden; cornfields near High Rocks and Langton Green; gravelly field on the Hertford road, near N.E. corner of Hatfield Park; fields between Woodstock Lane and Hookstreet (Sur.); Branfields, right of road from Talworth Toll to Talworth Court; between Cobham Common and Byesticle in a pasture field.

GENISTA ANGLICA.—Moors and heaths; frequent, 5–6. Wimbledon Common;* Hampstead Heath; Keston Common; Esher Common, &c.; Epping Forest.*

GENISTA TINCTORIA.—Pastures, thickets and borders of fields; not common; more frequent in Herts than elsewhere near London, 7–8. Box Hill (?); near Dorking, and on the Holmwood (*olim*); Keston Common (?); Warley Common (?); about Woodford, *olim* (?); meadow near Little Cheam; Rickmansworth Common Moor; near Verulam Buildings, St. Albans; near Scratch Wood, Edgwarebury; borders of Ball's Wood, Hertford;* lane leading to Brickendon Green from Broxbourne in meadows left-hand side;* meadows on western border of Wormley Wood;* near Wellington College; roadside, between Godstone and E. Grinstead; meadows near Abridge; pastures about Tunbridge Wells, frequent; Harrow Weald Common (?).

GENTIANA AMARELLA.—Pastures and old pits in the chalk districts, frequent, 7–9. Between Keston and Westerham; and in a chalky meadow two miles from Keston towards Down; old pits, Harefield;* about Northfleet; Box Hill;* Mickleham Downs;* in great profusion west of Wrotham on the slopes of the downs, on banks and about old chalk-pits;* Tring; Sawbridgworth.

GENTIANA CAMPESTRIS.—Rare, local. Colney Heath, among the furze in open spots;* No-man's-land Common, Herts; Waterdown Forest, Tunbridge Wells.

GENTIANA PNEUMONANTHE.—Moist, peaty heaths in Surrey, and in Tilgate Forest; Whitemoor Common, Worplesdon; Chobham Common,* plenty (near an isolated fir clump, left of the road from Chertsey); Woking Heath, E. of the cemetery; sparingly on Esher Common, in one place behind the first farmhouse on the road to Oxshott Hill; peaty sands near Bagshot (?); Waterdown Forest, Tunbridge Wells; in Tilgate Forest, between the Norfolk Arms and Balcombe; and in several places near Handcross.

GERANIUM COLUMBINUM.—Dry pastures in a gravelly soil (?) in the chalk districts, 6–7. Box Hill;* Reigate Hill;* Buckland Hill; Harefield;

Darent Wood;* Purfleet; about Hertford; St. Stephen's Hill; St. Albans; between Watford and Rickmansworth.

GERANIUM DISSECTUM.—Hedges, gravelly and waste places, common, 6–7. (Banks of the canal, Tottenham;* plenty.)

GERANIUM LUCIDUM.—Rocks and walls, mostly in hilly parts, not general, 5–8. About Godalming, frequent; lane leading from Harefield to the mill, bank between Barking and Dagenham; lanes between the Colne and the high road from Watford to Rickmansworth; wall on the road to Greenford.

GERANIUM MOLLE.—Dry pastures and waste places, common, 4–8. (Hampstead Heath,* upper parts, &c.)

Geranium phæum.—Rare, incidentally near habitations, 5–6. Thames Ditton; West Croome Park; roadside near Mickleham.*

GERANIUM PRATENSE.—Moist pastures and thickets, &c.; not common, 6–9. Borders of an osier holt, and by the ditch there between Mortlake and Kew* (converted, 1875, into a market-garden, but the Geranium may yet be extant by the ditch); foot of Box Hill and of Reigate Hill; Croydon (?); Eton meadows; Cookham; Reading; meadows near Holywell Bridge; St. Albans; meadow at Pinner; meadows &c., by the Thames, especially near Sunbury Lock; by the Mole at Esher and Stoke.

GERANIUM PYRENAICUM.—Meadows, pastures, roadsides, railway banks, frequent, 6–7. Roadside from Leatherhead to Mickleham;* Mitcham Common;* railway bank, Hampton Wick;* ditto near Kew by the river,* Hogs millstream, 1 or 2 miles from Kingston.

GERANIUM ROTUNDIFOLIUM.—Pastures and waste ground, rare, 6–7. Fields bordering Epping Forest.

GERANIUM ROBERTIANUM.—Woods, hedges, banks, stony and waste ground, common, 5–9. Everywhere in the outskirts of London in hedges.

GERANIUM PUSILLUM.—Waste ground and gravelly soils, frequent, 6–9. Everywhere in the outskirts, roadsides, canal and railway banks.*

GEUM URBANUM.—Hedges, common; frequent in the outskirts, in hedges, lanes, and ditch-banks.

GEUM RIVALE.—Very rare, 5–7. Nowhere nearer London than Royston, Herts, and banks of the Kennet at Newbury, Beds.; unless in meadows by the canal E. of Berkhampstead.

GLAUCIUM LUTEUM.—Sandy seashores, local, 6–10. Frequent about Southend.*

GLAUX MARITIMA.—Salt-marshes by the Thames below Woolwich, on both sides of the river, plentiful, 6–7. Abundant between Woolwich and Erith.*

GLYCERIA AQUATICA.—Sides of rivers, ditches and ponds; common, 7–8. (Plenty by the Thames* and Lea.*)

GLYCERIA FLUITANS.—Pools and ditches, common, 7–8. Everywhere in the outskirts.*

GNAPHALIUM SYLVATICUM.—Groves, thickets, and pastures; not common, 7–9. Charlton Wood, *olim* (?); Epping Forest: groves at Battleswell; near Harefield; Bagshot Heath;* Boar Hill woods; about Cold Harbour and foot of Leith Hill; Harrow Weald Common; Warley Common; Epping Forest (rarely); Godalming slopes towards Hurtmore; Redstone Hill; woods about Ranmer Common; near High Rocks; hilly field near Hoddesdon; St. Albans; gravel-pit by the road-side between Bayford and Little Berkhampstead; Harrow Weald Common; shady field near Harefield; Winter Downs; Fairmile Common; St. George's Hill; between Byfleet and Cobham.

GNAPHALIUM ULIGINOSUM.—Sandy and wet places by roadsides and on commons, frequent, 7–9. (Barnes Common; Wimbledon Common, &c.)

GNAPHALIUM DIOICUM.—Rare, 6–7. Banstead Downs.

GYMNADENIA CONOPSEA.—Dry pastures, and heaths in the chalk districts; not uncommon, 6–8. About Box Hill; between Croydon and Sanderstead; Purley Downs; Harefield; S. side of Albury Nowers Wood; about the Wray Pit; Red Hill; Farthing Downs, on a bordering bank;* Tring; Aston; Clinton and Ivinghoe, on all the chalk banks in those places. On Farthing Downs, with the ordinary form, a small white variety occurs sparingly.

GYMNADENIA ALBIDA.—Rare; Tilgate Forest.

HABENARIA CHLORANTHA.—Beech woods in the chalk districts, frequent, 5–8. Box Hill; woods and lanes near Farnham ('Botanical Gazette'); about Great Berkhampstead and several places in the Colne districts; Purfleet; Fyfield; wood between Essendon West-end and Kibes Green; Brickendon woods; Box Wood; Hertford Heath; Ball's Wood.

HABENARIA BIFOLIA.—Moist copses, &c., not frequent near London, 6–8. Darent Wood;* Box Hill and Betchworth; Norwood (?) *olim;* Charlton Wood (?); woods about Croham Hurst, between Down and Cudham; Farnham Wood; Epping Forest; near Sunninghill bog; Bisham Wood; Tring Heath (*olim*); hills E. of Merstham woods; left of road from Redstone Hill to Nutfield.

HABENARIA ALBIDA.—See GYMNADENIA.

HABENARIA VIRIDIS.—Dry hilly pastures in the chalk districts, rare, 6–8. About Coulsdon; Purley Downs;* Box Hill; near Tring, in chalky meadows at Barley End near Ashbridge; near Fyfield; Bedwell Park, Essendon; near Mount Pleasant, Brickendon; meadow near W. border of Wormley Wood.

HEDERA HELIX.—Woods, hedges, and old buildings; common, 10–11. Chingford Old Church, &c.*

HELIANTHEMUM VULGARE.—Dry pastures and waysides in the chalk districts; common, 7–9. In profusion on the Banstead Downs,* and warren by Epsom Downs* (growing in wide patches); frequent everywhere on the Downs, from the Hog's Back to Cuxton,* roadside from Dartford to Darent Wood;* ruins of Verulam, St. Albans; roadsides, Croydon to Selsdon and Sanderstead;* chalk-pits, Essendon.

HELLEBORUS FŒTIDUS.—Pastures and thickets, chiefly in the chalk districts; rare, 2–4. Wood left of Headley Lane; cliffs between Northfleet and Gravesend (?); Fridley Copse near Mickleham; hedge near High Laver Church; Purfleet, (?) *olim;* between Tring and the reservoirs near.

HELLEBORUS VIRIDIS.—Woods and thickets, chiefly on a chalky soil. About Coulsdon; wood near Chipstead Church; copse near Ranmore Common;

wood near Mickleham, left of the Dorking road; near Harefield; Barn Hill, near Harrow (?) *olim;* near Epping; in a wood near the railway goods station, Watford; Redhill; Lea districts, to wit, woods and bushy places on the chalk-pits at Watery Hall Farm, Hertingfordbury; meadows at Great Berkhampstead; between High Grounds and Hoddesdon; lane S.W. of Rickmansworth.

HELMINTHIA ECHIOIDES.—Borders of fields, chiefly in a clay soil; frequent, 6–10. Plenty on canal and railway banks, and by hedges in the environs.

HELOSCIADIUM INUNDATUM.—Pools and ditches that are dried up in summer; frequent, 7–8. (Putney Heath;* Wandsworth Common; Epping Forest;* Stanmore Marsh, &c.)

HELOSCIADIUM NODIFLORUM.—Boggy meadows, ditches, and watersides; common, 7–8. (Banks of the Thames;* Barnes Common;* ditches by the Lea, &c.*)

HERACLEUM SPHONDYLIUM.—Hedges and ditches, and on coarse pastures; common, 7. (Everywhere about London.)

HERMINIUM MONORCHIS.—Chalky pastures, rare, 6–7. Box Hill;* Reigate Hill, above the Wray chalk-pit; Purley Downs; chalk-pits on the Hog's Back; in Norbury Park; Albury Parsonage, Herts; Tring, with *Habenaria viridis*, q.v.; old chalk-pit, Morant's Court Hill; old chalk-pit, E. side of Hatfield Park; Ditton; Epsom Downs.

Hesperis matronalis.—Hilly pastures and copses, incidentally, 5–7. In a wood between Leatherhead and Dorking, below Box Hill; side of the Ravensbourne at Lewisham (?) *olim;* Sutton oil-mill, in a hedge near the Dock (?).

Hieracium maculatum.—Incidental on garden walls, 6–7. Highgate;* Teddington.*

HIERACIUM PILOSELLA.—Banks and dry pastures, common, 5–8. (Hampstead Heath;* Barnes Common.*)

HIERACIUM SYLVATICUM.—*See* H. VULGATUM.

HIERACIUM VULGATUM.—Woods, bushy heaths, commons and roadsides in wooded countries, frequent, 7–9. (Wimbledon Common and Putney Heath;* Hampstead Heath, among the fern;* &c.)

HIERACIUM TRIDENTATUM.—In similar situations, probably not unfrequent. (Is there really any essential difference between this and the preceding?) Bank on S. side of road leading to Whitton Park from Twickenham, above junction of loop-line; hedge bordering Fairmile Common; right of road ascending from Spaw Bottom; road leading from Bagshot to Staines, near Virginia Water.

HIERACIUM UMBELLATUM.—Woods, borders of woods and bushy heaths, common. (Plenty on Hampstead Heath.*)

HIERACIUM BOREALE.—In similar situations, 7–9. A mere variety of the preceding. Bank dividing Betchworth Park from the Dorking and Reigate roads; sandy banks between Dorking and Leith Hill; hedges bordering Sutton Common; Thieves Lane; Hertingfordbury; Hertford Heath; Bayford; Essendon; Wormley Wood; Northaw by the Ridgway; woods by Pinner Lane; Old Park Wood; Harefield; Harrow Weald Common; Stanmore Heath; Winchmore Hill Wood; about Esher; Cobham (Surrey),

St. George's Hill; Albury; Shiere; about Rickmansworth and Hertford; S.W. corner of Verulam ruins; St. Albans; Epping Forest; Warley Common; woods of Thorndon Hall; near Brentwood; Epping Forest.

HIERACIUM MURORUM.—Confounded possibly at times with *H. boreale* (?); in stony situations, not common, 6–8. Shiere; Castlewood; Shooter's Hill (?); woods and lanes between Dorking and Leith Hill; Cold Harbour Lane; Hortonwood between Ashtead Common and the course at Epsom; between Buckland and Buckland Hill; walls, Abbey Church, St. Albans, and other old walls about; Dorking chalk-pit;* Hurtwood Common, (confounded with *H. sylvaticum*, *i.e. vulgatum*, Cybele Brit.). [N.B. The student should collect these species from their several localities, and compare them with the descriptions in the books.]

HIPPOPHAE RHAMNOIDES.—Rare, local. Essex Coast near Canvey Island, 6–7.

HIPPOCREPIS COMOSA.—Banks and pastures in the chalk districts, frequent, 5–7. Box Hill;* Epsom Downs; lanes between Croydon and Sanderstead,* and between Croydon and Selsdon;* roadside between Dartford and Darent Wood;* between Northfleet and Gravesend; about Coulsdon; Reigate Hill in profusion,* old quarries W. of Dorking;* in many other places on the chalk downs,* and on the road from Cobham to Cuxton; Aldbury Nowers Wood; chalk-pit near Hertford.

HIPPURIS VULGARIS.—Ditches and stagnant waters, not of frequent occurrence, 6–7. Bogs on Uxbridge Moor, and several places in the Colne;* mill-pond near Leatherhead; pond at Bury Hill, near Dorking; pond below Moor Park; ponds, Clandon Park, Guildford; between Stanstead and Rye House; and in the Lea downwards; ditches in Hoddesdon Marsh, and by the railway at Waltham; reservoir, Tring; in the Lea, at Stanborough; pool at Walton Bridge.

HOLCUS MOLLIS.—Hedges and pastures, common. (Barnes Common;* Lea meadows, 7.*)

HOLCUS LANATUS.—Hedges, pastures and woods, common, 6–7. (Barnes Common;* Epping Forest, &c.*)

HONKENEYA PEPLOIDES.—Sea-shores local, 5–8. Southend; N. Shoebury.

HORDEUM PRATENSE.—Meadows and pastures, common, 6–7. (About Hendon;* Ditton Common.*)

HORDEUM MURINUM.—Waste ground by roadsides and walls, generally near towns and villages, common, 6–7. (Plentiful in the environs of London.*)

HORDEUM MARITIMUM.—On the banks of the Thames, below Purfleet and Erith, local, 6. Plentiful in the above localities, and adjoining marshy pastures.*

HORDEUM SYLVATICUM.—Woods and thickets, in a chalky soil, rare, 7–8. Berkhampstead; borders of woods, Tring;* between Maidenhead and Marlow; in shaws above fields behind the 'Fox,' Ranmore Common; Long Spring, Watford; Hillwood; Rickmansworth; Berkhampstead; Aldbury Nowers Wood; thicket by the roadside at River Hill, near Sevenoaks.

HOTTONIA PALUSTRIS.—Ditches and pools in Herts, Essex, and basin of the Thames, not unfrequent, 5–6. About Woking; Epping Forest, in pools between Walthamstow and Essex ;* between Weston Green and Ditton Common; between Greenwich and Woolwich *olim* (?); ditches between Ware and Hertford; ponds near 10th milestone, Romford road; Totteridge ponds; pools near the Welsh Harp, Hendon ;* in the Cran near Hounslow ;* Hoddesdon Marsh; near Watford; Harefield; Whitemoor Pond; Ash.

HUMULUS LUPULUS.—Thickets and hedges, frequent; especially in Kent, 7–8. Roadside between Crayford and Dartford; about the Crays; by the Ravensbourne; about Erith ;* frequent in a lane leading from Wrotham to Cuxton ;* &c.

HYACINTHUS.—*See* SCILLA NUTANS.

HYDROCHARIS MORSUS-RANÆ.—Ditches and ponds, principally near the Thames, 8. Backwater of the Thames at Walton ;* ditches in the marshy meadows below Higham ;* and also below Dartford ;* between Greenwich and Woolwich; and on towards Erith.* In the Paddington Canal ;* also in the canal between Harefield and Uxbridge ;* Barnes Common, in the ditch right, a patch of it ;* Rainham; Grays; Purfleet; Egham.

HYDROCOTYLE VULGARIS.—Boggy, and wet places on moors and heaths, common (plenty on Hampstead * and Putney Heaths *), 7–8.

HYOSCYAMUS NIGER.—Waste places, chiefly on a chalky soil, rare, and probably almost exterminated in the metropolitan districts, 6–8. About Hertford; Mead Lane; Ware Park; Hatfield Park; roadside near Ware; near Stanstead; Essendon; Sawbridgeworth; Box Moor; back road ascending from church Cobham to Fairmile; about Worplesdon; Pirbright; near the mill, Harefield (*olim*); Chislehurst Heath (?); near the palace, Eltham (?); on a common near Guildford (? Shalford, none there now); Abrook Common (?) *olim*; Red Hill and Reigate (? uncertain); Nutfield; Merrow Downs; Puttenham Heath and Common (possibly); common on the chalk about Farnham; Gomshall; on rubbish about St. Albans and Sandpit Lane. No-man's-land Common, Herts (more frequent, perhaps, towards Cambridgeshire); Fyfield Wood, near Grays; banks between Leigh and Southend, and Southend towards Shoebury; lane between Brockham and Godbroke; Cookham.

HYPERICUM ANDROSÆMUM. Hedges, and shrubby places, not very frequent, 6–8. Hills E. of Merstham; lane west end of Reigate Park; Leith Hill woods; lanes W. of Dorking ;* between Guildford and Dorking; about Claygate, Ditton; wood between Chislehurst and Bromley; thicket near Harefield Church (?); lane leading from Broxbourne to Brickendon Green ;* and other lanes in that neighbourhood; Ashbridge woods, and woods near Great Berkhampstead; wood near Thorndon; between Abridge and Romford; Cookham; Redstone Hill; copses on the Weald Clay, below Crawley, frequent. Weston Wood, Albury; Bookham; Stoke Wood.

HYPERICUM DUBIUM.—Rare, local, Hatfield.

HYPERICUM QUADRANGULARE.—*See* H. TETRAPTERUM.

HYPERICUM PERFORATUM.—Woods and thickets, common, 7–9. (Roadside from Putney Heath to Kingston, by the Park ;* lane leading from the 'Spaniards,' Hampstead Heath, to the meadows below Bishop's Wood,* &c.)

HYPERICUM TETRAPTERUM. Moist pastures, sides of ditches and rivulets, frequent, 7. (Ravine, Wimbledon Common ;* by the Lea ;* Broxbourne and Wormley woods.*)

HYPERICUM HUMIFUSUM.—Moory heaths, and commons, frequent, 7. (Wimbledon Common ;* Surrey heaths; generally in wet localities ;* Wandsworth Common.*)

HYPERICUM PULCHRUM.—Woods, and heathy places, frequent, 6–7. (Wimbledon Common ;* Hampstead Heath ;* Esher Common.*)

HYPERICUM HIRSUTUM.—Woods and thickets, frequent, and chiefly in a chalky soil, or on gravel over chalk, 7–8. Box Hill ;* and in thickets on the chalk slopes, generally from the Hog's Back to Cuxton ; as downs W. of Dorking ;* Hog's Back; hills E. of Merstham ;* hills E. of Shoreham,* and E. of Wrotham ;* Chislehurst Common ;* banks bordering Tring Woods ;* Broxbourne and Wormley woods ;* Burnham Beeches, and adjoining woods ;* Harefield ; Ball's Wood.*

HYPERICUM MONTANUM.—Bushy hills on the chalk range, not common, 7–8. Near Croydon; Riddlesdown, near Caterham Junction ;* near Warley ; Epping Forest (?) ; copses above Shiere ; hedgebanks towards Crown pits, Frith Hill, Godalming, plentiful ; Bisham Wood, Berks ; about Gatton, Nutwood, and Merstham ; Buckland Hills ; Norbury Vale ; Ranmore Common ; near Bussell's Green, Sevenoaks.

HYPERICUM ELODES.—Bogs, especially in the Surrey heaths, frequent, 7–8. Esher Common ;* Farnham Common ; by Burnham Beeches ;* bogs on Leith Hill ;* Bagshot Heath ;* Whitemoor Common ;* Pirbright Heath ;* boggy places by the canal beyond Woking ;* Witley Common ;* Puttenham Common ;* Keston Common ;* Putney Heath * (a patch), &c. ; Warley Common.

Hypericum calycinum.—Incidental, not frequent. Near the Dartford Powder Works ;* Mickleham, abundant ;* upper part of the park, Greenhithe ; road between Sutton and Banstead ; hills E. of Merstham ; W. end of Reigate Park ; Wotton ; Cold Harbour, Leith Hill.

HYPOCHŒRIS GLABRA.—Fields, and gravelly soils, rare, 6–8. Cultivated ground between Weybridge station and the town (?) *olim* ; in a field near Chobham towards Chertsey ; Wimbledon Common (?) ; Esher Common by the park palings, sparingly, towards Oxshott ; fields between Richmond and Kingston ; left of the road from Farnborough station to Frimley ; Tunbridge Wells Common ; Rusthall Common ; field next road on the footpath from Broadoak end towards Bramfield, Herts ; sandy fields between Witley and Milford.

HYPOCHŒRIS RADICATA.—Pastures and roadsides, common, 7. (Hampstead Heath,* &c.)

IBERIS AMARA.—Chalky fields, rare, 7. East end of Reigate Heath (?) *olim* ; cornfields in Herts and Essex, occasionally (Cybele Britannica). Royston ; Hoddesdon ; field N. of Pryor's Wood ; between Ware and Fanham's Hall.

ILEX AQUIFOLIUM.—Woods and hedges, frequent ; also in plantations, 5–8. Epping Forest ;* Holmwood ;* &c.

Impatiens fulva.—Banks of the Wey, and of the Grand Junction Canal between Harefield and Uxbridge,* &c., by the Thames at Hampton Court; local, abundant in its localities, 8–9.

IMPATIENS NOLI-ME-TANGERE.—Rare. Woods about Coulsdon (?) *olim.* near Guildford; in a wood at Chislehurst (?).

Impatiens parviflora.—Incidentally, rare, Bedwell; near Essendon.

INULA HELENIUM.—Moist pastures, rare, 7–8. Meadows near Harefield (?) *olim*; Gauntlet's Meadows, near Breakspeares, Harefield (?) *olim*; by a pond near Blue Close, Mangrove Lane, Hertford; sparingly in a field near Hook hamlet, through which is a footpath to Long Ditton Church.

INULA DYSENTERICA.—Ditches, and watery places, common, 7–9. (Everywhere in the environs of London.*)

INULA PULICARIA.—Moist, sandy, and damp commons, not unfrequent, 8–9. Ditton Marsh, by the roadside;* Colney Heath, near the village;* commons about Pirbright village;* Earlswood Common; Hatfield Heath; Brickendon Green; Lye Lane, Watford road, St. Albans; roadside between Hampton and Sunbury; about Ealing.

INULA CONYZA.—Banks and borders of copses, &c., in the chalk districts, 8–9. Box Hill;* Betchworth hills;* hills W. of Dorking;* and everywhere frequent along the chalk range, from the Hog's Back to Cuxton;* Epsom Downs;* Harefield;* Purfleet; Tring;* between Hatfield and St. Albans, but scarce.

INULA CRITHMOIDES.—Rare, local. In a salt-marsh, between Leigh and Southend, 7–8.

IRIS FŒTIDISSIMA.—Woods and thickets, mostly in a chalky soil; frequent in Kent; rare elsewhere, 5–7. Darent Wood;* woods about Gravesend and Rochester; wood at Purfleet; Epping Forest, about Woodford and Walthamstow (?) *olim*; Warley Common; woody places and hedges between Tunbridge and Sevenoaks; below the hills E. of Wrotham in one place;* woods, N. Mimms; beyond St. Albans, towards Dunstable, in roadside hedges; copse right of lane leading from Reigate Heath to Colley Hill; slopes towards Hurtmore, near Godalming; Bisham Wood; Welham Green; and woods at N. Mimms; pasture at Perivale.

IRIS PSEUDACORUS.—Ditches, and sides of rivers, frequent, 5–8. (Ditches by the Thames, near Mortlake;* by the Lea at Tottenham;* &c.)

Isatis tinctoria.—Cultivated fields, 7. Rare, incidentally for the most part. Guildford chalkpits;* and adjoining field (more abundant formerly, now rather scarce); banks of the Wey between Woodbridge and Stoke; in a field near Epping; near New Wandsworth station (Crystal Palace line).

JASIONE MONTANA.—Heaths and moors, common, 6–9. (Barnes Common,* Hampstead Heath,* &c.)

JUNCUS EFFUSUS.—Marshes, and ditches, common. Everywhere round London, in these localities.

JUNCUS CONGLOMERATUS.—Equally common with the above, in similar situations.

JUNCUS DIFFUSUS.—In marshes, rare. Darman's Green, Herts; near Cole Green, Herts, borders of Epping Forest; between Woodford and

Walthamstow; in a wood hard by Great Shrubbush; on a heath between Guildford and Woking stations.

JUNCUS COMMUNIS.—*See* J. EFFUSUS and J. CONGLOMERATUS, 7.

JUNCUS GLAUCUS.—Wet pastures, and by roadsides, common, 7. Everywhere round London, in suitable localities.

JUNCUS MARITIMUS.—In salt-marshes, local, 7–8. Southend; about Tilbury Fort.*

JUNCUS ACUTIFLORUS.—Bogs &c., common. (Reigate Heath; Hampstead Heath; pond near right of lane leading to Fortune Green.*)

JUNCUS LAMPROCARPUS.—Boggy ground and marshes, frequent, 7–8. (Barnes Common;* Putney Heath.*)

JUNCUS OBTUSIFLORUS.—Wet pastures and marshes, not unfrequent, 8. (Hampstead Heath;* Putney Heath;* Tilbury Marshes; Reigate Heath;* Leith Hill.*)

JUNCUS SUPINUS.—Bogs and swamps, frequent, 6–8. (Hampstead Heath;* Putney Heath.*)

JUNCUS COMPRESSUS.—Marshes, common, 6–8. (Putney Heath;* Wimbledon Common;* Rainham; Tilbury.)

JUNCUS BUFONIUS.—Marshes &c., common, 6–7. (Hampstead Heath;* Putney Heath.*)

JUNCUS SQUARROSUS.—Heaths and moors, common, 6–7. (Putney Heath;* abundant.)

JUNIPERUS COMMUNIS.—Surrey and Kentish chalk downs, frequent, and in many localities on the range from Dorking to Cuxton, 5–6. Box Hill and the Betchworth hills;* and hills W. of Derking;* between Betchworth and Reigate, and on Reigate Hill;* hills E. of Merstham;* E. of Wrotham* and of Shoreham;* Banstead Downs;* Aldbury.

KNAUTIA ARVENSIS.—*See* SCABIOSA ARVENSIS.

KŒLERIA CRISTATA.—Pastures chiefly on the chalk, not common, 6–7, Reigate Hill;* hills E. of Merstham; Blackheath (?) *olim*; Kenley Common, plentiful;* roadside near Staines; commons near Southend; Tring; between Hertford and Ware; gravel-pits, Moulsey Hurst; Brockham Hill; downs about Shiere.

KONIGA.—*See* ALYSSUM MARITIMUM.

LACTUCA VIROSA.—Banks and waysides, frequent, 7–8. About the railway bridge near Willesden Junction, Acton road;* between Willesden and Neasdon;* between Neasdon and the Edgware road; Southend and Canvey Island;* Pinner road, Greenford; Pinner woods.

LACTUCA SCARIOLA.—Waste ground, rare. Lane by Charlton Church, and sand-pit in Charlton Wood (?) *olim*; about Croydon (?) *olim*; Southend, towards Leigh.

LACTUCA SALIGNA.—Waste ground near the Thames, and in salt-marshes, rare. Thames bank, near Erith;* near Leigh, towards Southend.*

LACTUCA MURALIS.—Old walls and in woods, not unfrequent, 6–8. Wall by the 'Spaniards,' Hampstead Heath;* Tring, beechwoods, plenty;* beechwood at Shoreham;* bank, left, near Sanderstead;* about Dorking, between Claremont and Ditton Common; and by the road from Kingston Common to Long Ditton Church; lane from Harefield to Ruislip from Iver to Fulmer.*

LAMIUM ALBUM.—Hedges and waste places, common, 5–9. Everywhe in the environs.

LAMIUM AMPLEXICAULE.—Cornfields and waste places, common. (Plentifi in cornfields beyond Croydon,* and between Richmond and Kingston.*)

Lamium maculatum.—Incidentally near cottages; not frequent, 6–8.

LAMIUM INCISUM.—Cultivated and waste ground, common, 4–6. Fr quent in market gardens and waste places in the environs.*

LAMIUM PURPUREUM.—In similar situations; common, 4–10. Every where in the environs.

LAMIUM GALEOBDOLON.— Woods, frequent, 4–6. Woods about Croydon; Ditton; Charlton Wood; Harefield;* Darent Wood;* Coulsdon; Ball Wood,* &c.

LAPSANA COMMUNIS.—Hedges, &c., common, 6–8. Hedges everywher in the environs. (Brondesbury; Willesden;* Mortlake.*)

LATHRÆA SQUAMARIA.—Woods and coppices, rare, 3–5. About Coul don; Mickleham; Guildford; Westhumble Lane (leading from Burfor Bridge to Ranmer Common),* field between Shalford Turnpike an Chantry Downs; lane leading from Harefield to the river; woods a Chipstead and Upper Gatton; wood S. of Chislehurst Common; hedge rows W. of Dorking; Great Berkhampstead woods; wood near No-man's land; Hazel copse, Woodland, near Farningham.

LATHYRUS APHACA.—Borders of sandy and gravelly fields, rare, 5–8 Cornfields at Ongar; Broomfield (?); Tottenham, and Enfield (?) Croydon (?); fields near Norton Heath; Southend; near Hoddesdon Sawbridgeworth.

LATHYRUS NISSOLIA.—Bushy places and grassy borders of fields, rare, 5–6 Fields near Chingford; Broomfield; Harefield, near the church; abou Tottenham (?); bank at Purfleet (?); roadside near Chislehurst (?) roadside near Finchley* (coming from Hampstead); Mitcham Common bank between Merstham and Wray Common; about Charlwood; Fel bridge; Brentwood; bank bordering Gatton Park; Southend; severa places about Hertford, as between Ball's Wood and Hertford Heath meadow by Tottenham High Cross; field on Colley farm, Reigate an elsewhere about Reigate.

LATHYRUS HIRSUTUS.—Cultivated fields, rare, 6–7. Cornfields at War lingham, below Wormsheath, 6 miles from Croydon (bank above) bushes below Hadleigh Castle; near Rawreth church (Essex Flora) Nasing, E. of Broxbourne.

LATHYRUS PRATENSIS.—Moist meadows, pastures, and hedges, frequent 7–8. (Plenty of the meadows below the 'Spaniards' towards Finchley.*

LATHYRUS SYLVESTRIS.—Thickets and hedges, not common, 6–8. Abou Coulsdon; Wickham Wood; Merstham pools, and adjoining bank; rail way bank near the tunnel, Merstham;* woods at Erith, and betwee Rochester and Northfleet; borders of Cobham Park;* chalky detritu at the foot of White Hill, E. of Merstham, in profusion* (E. de C.).

Lathyrus tuberosus.—Local, rare, 7. Near Southend; Canvey Island, in the grey marsh; plentiful about Fyfield, in cornfields and bordering hedges.

LATHYRUS PALUSTRIS.—Boggy meadows, rare, 6–8. At Peckham formerly (Ray).

LAVATERA ARBOREA.—Local, rare, 7–9. Southend Cliffs, above.

LEERSIA ORYZOIDES.—Banks of rivers, &c., rare, local (?), 8–10. In the Mole river, at Brockham Bridge, and other places; also between E. Moulsey church and Ember Bridge (Cybele Brit.); by the bridge over the canal near Woking station.

LEMNA MINOR.—Stagnant water, common, 7. Everywhere in the suburbs.

LEMNA GIBBA.—Clear stagnant water, rare, 6–9. Farnham (Bot. Gazette); ditches about Rainham; Epping Forest, near Leytonstone.

LEMNA TRISULCA.—Clear stagnant waters, rare. Rickmansworth; meadows at Ponder's End, by the path to Chingford;* ditches below Higham.*

LEMNA POLYRRHIZA.—Stagnant waters, rare; Epping Forest, near Woodford, in pools and holes;* Warley Common; about Ditton; Chertsey; Gatton pond; about Guildford; Reigate; Dorking.

LEMNA ARRHIZA.—*See* WOLFFIA.

LEONTODON HIRTUS.—Moors and gravelly pastures, frequent, 7–8. (Barnes Common;* Wimbledon Common.*)

LEONTODON HISPIDUS.—Meadows, pastures, and gravelly heaths, frequent, 6–9. (Wandsworth Common;* Hampstead Heath.*)

LEONTODON AUTUMNALIS.—Meadows and pastures, common, 8. (Hampstead Heath;* pastures everywhere.*)

Leonurus cardiaca.—Incidental, hedges and waste places rare, 7–9. Near Croydon, *olim* (?). Coulsdon (?); 4 m. from Godalming, towards Haselmere.

LEPIDIUM RUDERALE.—Waste places near the sea and on rubbish; frequent, 5–8. By the Thames, near Erith;* Canvey Island;* about Hampstead, West-end;* Gravesend.

LEPIDIUM LATIFOLIUM.—Marshes near the sea. Purfleet (?) *olim;* marshes at Grays (?); Barking.

LEPIDIUM CAMPESTRE.—Cornfields and dry gravelly soil, frequent, 5–8. About Croydon;* Caterham Junction;* Edgware road, beyond Brondesbury;* West Drayton; Epping; Grays; Purfleet, &c.

LEPIDIUM SMITHII.—Borders of fields and hedges, rare, 4–8. Not uncommon about Reigate, Box Hill, and Betchworth; Chislehurst; Charlton, Mids.

Lepidium Draba.—Incidentally, rare, 5–6; a patch of it by the Thames, between Hammersmith and Putney* (E. de C.) (? from seed washed up from Thanet); Barking.

Lepidium sativum.—Incidentally, near cottages; a garden waif, 5–6.

LEPTURUS FILIFORMIS.—Local, by the Thames, in several places, not common, 7. Near Erith;* Purfleet (var. *incurvatus*);* Tilbury Fort, in a marshy spot E. of it;* by the canal below Gravesend, in profusion a mile or so beyond the town.* (E. de C.)

LEUCOJUM ÆSTIVUM.—Local, rare, extinct, 5–6. By the Thames (*olim*) (?), on both sides of the river; Windsor; (in the former locality, *Allium triquetrum* mistaken for it: H. C. Watson, 'Topog. Bot.').

LIGUSTRUM VULGARE.—Thickets and hedges, frequent, 5–6. Hedges and enclosures, often in the suburbs of London, but planted (?); in thickets on the chalk.*

Lilium Martagon.—Incidentally, rare. Totteridge Park; avenue at Cobham* (planted, but now wild).

LIMNANTHEMUM NYMPHÆOIDES.—Still waters and rivers, rare, 7–8. In the Thames at Walton;* at Kingston;* in the Roding at Woodford (?) (*olim*); and above Ongar (*olim*); Thames at Cookham and at Sunbury; canal by the railway, New Cross.

LIMOSELLA AQUATICA.—Muddy borders of ponds, and where water has stood, rare; perhaps often overlooked, 7–9. About Croydon, plentifully; Coulsdon; warren pond, Breakspeares; Harefield; pond on Ditton Common;* on Shalford Common; pond at Finchley (?) *olim*; Elstree reservoir; on a riding in North Mimms Wood; ditch on Holmwood Common; ditches, Walton-on-the-Hill; Clapham Common (?)

Linaria Cymbalaria.—On walls, frequent, 5–9. On walls in many places in the environs. (Hampstead;* Highgate,* &c.)

LINARIA ELATINE.—Cornfields in a gravelly or chalky soil, not general, 7–10. About Epsom;* Thames Ditton parish (cornfield below Oxshott Hill);* Woodford; Chigwell; Broomfield; cornfields in several places on the chalk range, as Reigate Hill,* and W. of it;* cornfields below hills W. of Dorking;* Coulsdon.

LINARIA SPURIA.—Sandy cornfields, frequent, 7–11. About Dorking;* cornfields below the hills W. of Reigate, in profusion;* about Farnborough; Bloomfield; Harefield; field near Hurlesdon;* about Dartford; Croydon;* Guildford.

LINARIA REPENS.—Chalky banks, rare; 7–9. Near Gravesend; hedge near Sefton Arms, Stoke (Bucks.); garden-wall, Dagnall Lane, St. Albans.

LINARIA MINOR.—Sandy and chalky fields; frequent in the chalk districts, chalk-pits, &c. Northfleet and Greenhithe; chalk-pit, foot of the Addington hills;* chalk-pits, Harefield;* in cornfields on Reigate Hill;* and in cornfields, foot of the hills westward.*

LINARIA VULGARIS.—Hedges and borders of fields, common, 7–10. Everywhere about London.*

Linum usitatissimum.—Incidentally in cornfields, 7. Near Finchley Road railway station, a few plants;* (perhaps often about railway stations and oil-mills).

LINUM ANGUSTIFOLIUM.—Sandy and chalky pastures, &c., near the sea, rare. Bank between Leigh and Southend,* 5–9; near Cuxton, on the road to Stroud ('Bot. Gazette'); near Purfleet; foot of Box Hill; fourth field from Chessington church, through which is a footpath to Horton and Epsom.

LINUM CATHARTICUM.—Pastures, common, 6–9; abundant everywhere on the downs; Box Hill;* Reigate Hill,* &c.; Pinner Wood, borders;* Tring,* &c.; Dartford;* hills E. of Merstham and of Shoreham,* &c.

LITHOSPERMUM ARVENSE.—Cornfields and waste ground, especially in the chalk districts, frequent, 5–6. About Croydon; Greenhithe, towards Darent Wood;* Box Hill; Harefield; cornfields about St. Albans; between Cobham and Cuxton; common, S. of Hertford; Hoddesdon, by the new mill; fields near Oster Mills, St. Albans; Sawbridgeworth.

LITHOSPERMUM OFFICINALE.—Dry, waste, uncultivated places; borders of hedges and woods. Not unfrequent on the chalk near Darent Wood, road from Greenhithe;* about Dartford; Northfleet; Gravesend; between Down and Cudham; Chislehurst;* (a locality here is in the hollow by the railway crossing, below wood S. of the common, east;) Purfleet; Long Valley Wood, Rickmansworth; Fyfield; Hatfield Forest; hedge at Welstone, near Tring; Harefield; foot of hills E. of Merstham; slope of Ranmore hills towards Effingham; hills near Bletchingley; copses about West Humble Lane; Burford Bridge.

LITHOSPERMUM PURPUREO-CÆRULEUM.—Local, rare, 6–7. Darent Wood, towards Greenhithe.

LITTORELLA LACUSTRIS.—Borders of ponds and canals in a gravelly soil. Abundant in the Surrey Canal, about Woking;* ponds at Frensham; Esher Common (?) *olim*; Ruislip reservoir; Langley Heath, Bucks.; Windsor Park; about Reading; near Ascot; Pirbright.

LISTERA OVATA.—Woods, frequent, 5–7. Pinner and Oxhey woods;* Darent Wood, abundant;* copse near Harefield on the Uxbridge road; Box Hill and the Betchworth hills; Keston Common; Epping Forest; wood W. of Warley Common; Ruislip woods; woods about Croham Hurst, &c.

LOLIUM PERENNE.—Pastures and waysides, common, 6–7. Everywhere.*

LOLIUM TEMULENTUM.—Cornfields, frequent, 6–7.

LONICERA PERICLYMENUM.—Hedges and thickets, common, 6–9. Hedges everywhere in the environs a little way out;* Epping Forest,* &c.

Lonicera Xylosteum.—Rare, incidental, 5–6. Hedge near Brook, Godalming.

LOTUS CORNICULATUS.—Pastures, common. (Barnes Common;* Hampstead Heath.*)

LOTUS MAJOR.—Sides of ditches and watery places on heaths, frequent, 7–8. (Barnes Common;* Hampstead Heath, in the bog.*)

LOTUS TENUIS.—Banks, &c., not common, 7–8. Bank by the road from Dartford to Greenstreet Green.*

LUZULA SYLVATICA.—Woods in hilly districts, frequent, 5–6. Darent Wood;* Epping Forest;* Warley Woods; wood below Headley church, towards Walton; Wormley Wood; Pinner Wood; Tilgate Forest.

LUZULA FOSTERI.—Groves and thickets in a gravelly or chalky soil, rare; ex. in Surrey, 3–6. Left side of Telegraph Hill, near Ditton; about Dorking; about Dartford, and between Dartford and Crayford; Coulsdon; Hainault Forest; Pinner Wood; Harrow Weald Common (?) (*olim*); dry banks about Farnham; Bisham Wood; S. end of Easney Park Wood; woods by Pinner Lane.

LUZULA PILOSA.—Woods, frequent, 3–5. (Epping Forest;* Pinner Wood.*)

LUZULA CAMPESTRIS.—Woods and dry pastures, common, 4–5. (Hampstead Heath.*)

LUZULA MULTIFLORA.—In similar situations, and as frequent, 4–5. (Putney Heath and Wimbledon.*)

LYCHNIS FLOS-CUCULI.—Moist meadows, &c., frequent, 5–6. (Hampstead Heath, in the bog.*)

LYCHNIS VESPERTINA.—Under hedges and in grass fields, frequent, 6–9. Hedges in the environs;* (roadside between Barnes and Mortlake, &c.*)

LYCHNIS DIURNA.—Hedgebanks and in copses, frequent. Hedges in the environs; (Tottenham;* Hendon;* Willesden, &c.*)

LYCHNIS GITHAGO.—Cornfields, frequent, 6–8. Cornfields everywhere; if not in one, in another.

LYCOPSIS ARVENSIS.—*See* ANCHUSA ARVENSIS.

LYCOPUS EUROPÆUS.—Ditches and river banks, common, 6–9. (Plenty by the Thames, Lea, &c.*)

LYSIMACHIA VULGARIS.—Sides of rivers, and in wet shady places, frequent, 7–8. Hedges in the wet meadows by the canal between Harefield and Uxbridge;* by the Brent in two or three localities;* in the Wey, near Byfleet;* Guildford; banks of the Roding;* of the Lea; hedges about Felbridge; near Pirbright Common;* by the Thames at Teddington; at Mortlake *olim* (?); by the rivulet above the pond near Puttenham.*

LYSIMACHIA NUMMULARIA.—Damp, shady woods, banks, and pastures, 6–7. Darent Wood, abundant;* Chislehurst;* Keston Common; Epping Forest;* by the Thames in several places above Richmond;* Burnham Beeches;* by the canal, &c., Rickmansworth;* ditches in lanes about Colney Heath,* &c.

LYSIMACHIA NEMORUM.—Shady woods, frequent, 7–9. Darent Wood;* Pinner Wood;* Croham Hurst;* Box Hill;* Burnham Beeches;* Wimbledon Common;* Epping Forest; Warley woods; Broxbourne and Wormley woods.*

LYTHRUM SALICARIA.—River-sides and marshy places, common, 7–9. (Plenty by the Thames,* Lea, &c.*)

LYTHRUM HYSSOPIFOLIA.—Moist places, rare; (?) perhaps often overlooked, 6–10. Colney Heath, and in an adjoining field (?) *olim*; Hounslow Heath (?) *olim*, possibly by the Cran; between Staines and Laleham, in a marshy field (?) *olim*; by the Wandsworth steamboat pier in 1855 and 1859.

MAIANTHEMUM BIFOLIUM.—*See* SMILACINA.

MALAXIS PALUDOSA.—Sphagnum bogs, 7–9, very rare. Bog, corner of the pond on Puttenham Common, nearest Hampton Lodge; bog, foot of Oxshott Hill, a few plants, *olim*; bog on common adjoining Burnham Beeches (?); Ashdown Forest, near Tunbridge Wells, in the great bog by Kidbrooke Park pales; and at Pressbridge Warren, near Wychcross.

MALVA SYLVESTRIS.—Waste places, common, 6–9. Everywhere in the outskirts.*

MALVA ROTUNDIFOLIA.—Waste places, common, 6–9. Frequent in the outskirts;* (banks of the Lea Canal, plenty.*)

MALVA MOSCHATA.—Meadows, pastures, and roadsides in a gravelly soil, frequent, 7–8. About Merstham; Nutfield;* Bletchingley; Shiere; Albury; Guildford; Clammer Hill, near Godalming; Reigate Heath,* near Colney Heath,* &c.*

MARRUBIUM VULGARE.—Waste places, chiefly in the chalk districts. Hayes Common,* Box Hill, and Betchworth; Hog's Back,* Chislehurst;

Thursley; Reigate Heath;* Uxbridge Moor; Hampstead Heath, two plants;* Wimbledon and Streatham commons, *olim;* Tilbury?

Matricaria Parthenium.—Hedgebanks and waste places, generally near habitations, not unfrequent. About Hendon, and the suburbs of London, in several places.

MATRICARIA INODORA.—Fields, wastes, and waysides, common, 7–9. Everywhere, in waste places about the suburbs.

MATRICARIA CHAMOMILLA.—Waste places and cultivated grounds, frequent, 6–8. With the preceding, in similar situations and in market gardens.*

Medicago sativa.—In fields, hedges, and borders of fields, frequent, 6–7. Cultivated fields in the environs, and bordering hedges; near Mitcham.*

MEDICAGO MACULATA.—Gravelly pastures, frequent, 5–8. (Putney Heath, plentiful.*)

MEDICAGO LUPULINA.—Waste and cultivated ground, frequent, 5–8. Ditch banks by the Thames near Erith, &c., plentiful,* &c.

MEDICAGO DENTICULATA.—Rare, 5–8; local. Behind Ponder's End, *olim.*

MEDICAGO FALCATA.—Dry gravelly banks, rare, 6–7. About Croydon (?) *olim*; between Watford and Bushey Hill (?) *olim.*

MEDICAGO MINIMA.—Sandy fields and wastes, rare; local, near Southend, 6.

MELAMPYRUM PRATENSE.—Woods and thickets, frequent, 5–8. Pinner and Ruislip woods;* Darent Wood;* copses, Highgate;* Epping Forest;* Croham Hurst;* Box Hill.*

MELAMPYRUM CRISTATUM.—Rare, 7. In newly-cut copses on chalk, N.W. of Hertford; N.W. corner of Essex in similar localities; Box Hill, or wood near Headley (*olim*), doubtful; wood near Northchurch Common; Tring, *olim* (?).

MELAMPYRUM ARVENSE.—Cornfields, rare, 6–8. In Herts, near Ashwell.

MELICA UNIFLORA.—Shady woods, frequent, 5–7. Hedges about Hendon,* &c.

MELILOTUS OFFICINALIS.—Waysides and bushy places; borders of fields, frequent. Banks of the Thames;* Harrow Road;* about Box Hill; Dorking; Epping Forest.

MELILOTUS ALBA.—Waysides and bushy places, occasionally, not very frequent, 7–8. About Erith;* marshy meadow near Cuxton railway station, abundant;* Mickleham (*olim*); about railway stations occasionally.

Melilotus arvensis.—Rare, local, 7–8. 'Cybele,' iii. p. 332.

MELITTIS MELISSOPHYLLUM.—About Furnace Pool, Felbridge, in great variety (*olim*), (doubtful; Cybele Brit.) local, rare.

Melissa officinalis.—A casual. Roadside between Chelmsford and Galleywood Common; private road from Chessington to Leatherhead.

MENTHA ROTUNDIFOLIA.—Moist places, rare, 8–9. Wimbledon Common (?); Woolwich marshes (?) (now drained); between Crayford and Dartford (?); by the Ravensbourne at Lewisham, (?) *olim;* about Harefield; lane between Reigate church and Ffrenches (?); by the roadside between Hertford and Essendon, opposite Watery Hall Farm; in a blind lane a little off the new road between Bedwell and Camfield Place,

Essendon; lane between Pembridge and Westuble lanes; Worplesdon; stream near Hooley House, Redhill; Ashtead Park, near Epsom, by the footpath.

MENTHA SYLVESTRIS.—Moist waste ground, not common, 8–9. About Dartford; between Crayford and Dartford; between Harlow and Sawbridgworth; Epping Forest; between Camfield Place and Hatfield Woodside; Hoddesdon, by the railway (mill stream); near Emmett's Mill, between Ottershaw Park and Chobham.

Mentha viridis.—Marshy places, doubtfully wild, near London, rare, 8. Side of a pond, in a field 2 miles from Thorndon (*olim*); banks of the Thames (?) (where ?)

MENTHA PIPERITA.—Watery places, rare, 8–9. Cultivated about Mitcham, and on the Sutton Downs; Croydon, by the river; (waif of cultivation ?) about Guildford.

MENTHA AQUATICA.—Marshes, ditches, and banks of rivers, frequent, 8–9. In such-like localities everywhere round London.

MENTHA HIRSUTA.—A mere variety.

MENTHA CILIATA.—Ditto.

MENTHA SATIVA.—Wet places, banks of rivers, ditches, &c., frequent, 7–8. In ditches by the Lea;* Colney Heath, and near it;* Epping Forest.*

MENTHA RUBRA.—In similar situations, 8–9, rare, a mere variety. Ditches, near Purfleet;* between Ongar and Brentwood; Sonning, Berks; Nutfield Marsh.

MENTHA ARVENSIS.—Cornfields, frequent; near Harefield;* cornfields foot of the hills E. of Dorking, plentiful;* cornfield, foot of Oxshott Hill.*

MENTHA GENTILIS.—A variety of the preceding; in similar situations. By a mill-stream at West Ham.

MENTHA PRATENSIS.—Watery places, in moist meadows, not frequent, 8–9. Pirbright Common, near Worplesdon, in the hollow;* by the Lea.

MENTHA GRACILIS.—A variety of the last; in similar situations, 8–9.

MENTHA PULEGIUM.—Wet commons and margins of brooks, not frequent. Ditton Marsh;* Mitcham Common;* Wimbledon Common (?) *olim*; Wandsworth Common; gravel-pits in Epping Forest; about Coulsdon; Dartford, by the Cray; Whitemoor Common;* Reigate Heath; Harefield; Coulsdon; Purfleet; Chislehurst Common; Earlswood Common; Totteridge Green; Colney Heath;* Brickendon Green.

MENYANTHES TRIFOLIATA.—Marshes and bogs, frequent, 5–7. Wimbledon Common, ravine near the mill;* Hampstead Heath;* Hayes Common;* bog between Farnborough and Keston Common; Reigate Heath;* Felbridge pools; Colney Heath;* Rickmansworth Moor; in Epping Forest; Warley Common; bogs, Waterdown Forest, Tunbridge Wells; Colney Heath;* Box Moor, Berkhampstead; Ball's Wood; Harefield Moor.

MERCURIALIS PERENNIS.—Woods and shady places, common, 3–5. Shady lanes and groves everywhere in the environs.

MERCURIALIS ANNUA.—Waste places about habitations, not common, 7–11. At Child's Hill, in cottage plots, &c.;* Croydon; elsewhere about London; banks of the Thames; Southend.

MESPILUS GERMANICA.—Hedges, rare, 5–6. Redhill; between Redhill and Nutfield.

MILIUM EFFUSUM.—Moist shady woods, frequent, 5–6. Lane from Reigate Hill to Colley Hill; Epping Forest; Charlton Wood; copses near Reigate; Brondesbury Park;* Forest Hill.

Mimulus luteus.—Rare, incidental. Bourne Bridge, between Woking and Chobham.

MŒNCHIA ERECTA.—Gravelly pastures, frequent, 5–6. Wimbledon Common;* Barnes Common;* Moulsey Hurst;* Epping Forest; Reigate Heath; Albury and Shiere heaths; Harefield; Uxbridge Common; Warley Common; Colney and Hertford heaths.

MOLINIA CÆRULEA.—Moors and wet heaths, frequent, 7–8. (Putney Heath;* Hampstead Heath.*)

MONOTROPA HYPOPITYS.—Dry beech and fir woods, not common, 7–8. Box Hill;* Mickleham, in the wood left of Headley Lane, further end especially;* Reigate Hill;* Coulsdon; Tring; Shiere.

MONTIA FONTANA.—Rills and about springs, frequent, 4–8. (Hampstead Heath;* Barnes Common;* Ditton Marsh, by the railway.*)

MYOSOTIS PALUSTRIS.—Ditches and river-sides, common, 6–8. (By the Thames;* Lea;* Colne;* and especially in bordering ditches; Wimbledon Common, &c.)

MYOSOTIS CÆSPITOSA.—Watery places, common, 6–8. (Wimbledon Common;* Putney Heath.*)

MYOSOTIS REPENS.—Peat bogs, rare, 7–8. Peat bogs about Farnham; Wimbledon Common; Bagshot Heath; bog, Little Berkhampstead, Herts; Bell Bar bog; Hatfield woodside; Ruislip reservoir; ditches near Pinner Drive; about Esher; near Gracious Pond; boggy spots on Leith Hill; near Witley station, in a bog.

MYOSOTIS SYLVATICA.—Dry, shady places, and in dry woods, not common, 5–8. Pinner and Oxhey woods;* Darent Wood;* roadside between Crayford and Dartford; Epping Forest; wood on Tilsey Hill, 8 miles from Croydon (borders).

MYOSOTIS ARVENSIS.—Cultivated fields and hedgebanks, common. (Fields between Willesden and Harlesden Green);* and elsewhere in the environs.*

MYOSOTIS COLLINA.—Dry, sandy banks, not unfrequent, 4–5. Barnes Common;* Wimbledon Common;* Sheen Common,* &c.

MYOSOTIS VERSICOLOR.—In dry and wet situations, frequent, 5–6. Barnes Common;* Shirley Common;* bank by the roadsides about Hounslow, Teddington, &c., plentiful.*

MYOSURUS MINIMUS.—Cornfields and waste places, chiefly in a chalky soil, rather uncommon. Epping Forest (borders?); Warley Common (?) *olim*; cliffs between Northfleet and Gravesend; Claygate, near Ditton; Epsom; west end of Reigate Park; cornfields near Frensham Church; about Hertford; cornfields, Slough; Cookham, Berks; in the closes at Streatham; (?) fields near Cæsar's Camp, Wimbledon; Fflanchford and Santon cornfields; several places between Hertford and Hatfield; between Hoddesdon and Rye House; gravelly fields, North Mimms; wheatfield, west end of Park Hill, Reigate.

MYRRHIS ODORATA.—Rare, local; side of road between Upper Gatton and Shabden Park, Chipstead.

MYRICA GALE.—Bogs and moors, local, 5–7. Bagshot Heath* (left of the road from Chobham); Pirbright Heath;* Bisley Common; neighbourhood of Stroud House, near Godalming; Broadmoor Bottom, Berks; Waterdown Forest, near Tunbridge Wells.

MYRIOPHYLLUM VERTICILLATUM.—Ponds and ditches, not unfrequent; pond on the common by Walton Bridge, where the Limnanthemum grows;* in the Roding at Woodford; in the Brent at Greenford; marsh ditches below Greenwich (?) in the Colne ditches about Rickmansworth and Uxbridge; between West Ham and the Thames; pond, Warley Common; Fyfield; ditches by the railway, near confluence of the Lea and Stort; dam near old mill, Hoddesdon; Tring reservoir; above Weybridge, in marsh near the river.

MYRIOPHYLLUM SPICATUM.—Ditches and stagnant waters, frequent, 6–7. Wimbledon Common; ditches in the marshes below Woolwich, by the Thames, abundant;* in the Roding at Woodford; round pond, Bushy Park; Colney Heath; ponds on Highdown Heath; Earlswood Common; Felbridge pools.

MYRIOPHYLLUM ALTERNIFOLIUM.—Rare; in ponds. Ditton Marsh; Abrook Common; Frensham little pond;* gravel-pits, Higham Bushes; Loughton; pond at Brickendon Green; Totteridge Green; Colney Green; Elstree reservoir; Great Berkhampstead Common; canal at Ash.

NARCISSUS PSEUDO-NARCISSUS.—Moist woods, thickets, and pastures, not frequent, 3–4; Pinner;* Charlton Wood (?) *olim*; orchard at Breakspeares, near Harefield; near Croham Hurst; field at Low Layton (?) *olim*; meadow on a farm near Woodhatch; copse right of lane leading from foot of Redstone Hill to Nutfield; Mill Hill; Hookwood Wood, near Worplesdon.

Narcissus major.—Rare, incidental; a garden waif.

Narcissus biflorus.—Incidental. On Totteridge Green (*olim*); and also in several places near Harefield (*olim*); meadow S. of Ruislip reservoir; Tilgate Forest.

NARDUS STRICTA.—Wet heaths and moors, common. (Hampstead Heath;* Putney Heath.*)

NARTHECIUM OSSIFRAGUM.—Bogs on heaths and moors, frequent, 5–8. Keston Common;* Bagshot Heath;* Pirbright Heath;* Esher Common;* Farnham Common by Burnham Beeches;* bogs on Leith Hill;* Putney Heath; a small patch near entrance to Roehampton lane;* Chislehurst.

NASTURTIUM OFFICINALE.—Brooks and rivulets, common. Hampstead Heath, sparingly, but frequent in the outskirts; by the Thames; bordering ditches by the Lea, &c.*

NASTURTIUM SYLVESTRE.—Watersides and and waste places, frequent, 6–8. Plentiful, roadside between Hammersmith and Barnes Common;* roadside by the lake, on the embankment, &c., at the Welsh Harp, Hendon.*

NASTURTIUM TERRESTRE.—Watery places, not very common, 6–10. Thames side between Putney and Kew; borders of a pond right of the lane from Hampstead Heath to Fortune Green.*

NASTURTIUM AMPHIBIUM.—Watery places, frequent, 6–8. (By the Thames between Putney and Mortlake;* ditch on Barnes Common.*)

NEOTTIA NIDUS-AVIS.—Shady woods, frequent, but not in any abundance, 5–6. Charlton Wood (?) *olim*; White Heathwood, near Harefield; Epping Forest, below Woodford;* Nutfield copse, by the Wray Common; Reigate Hill;* Box Hill; woods about Coulsdon; in a wood near Felbridge; Fridley copse, Mickleham; Beeching Wood, Norbury Park; Darent Wood;* woods about Ranmore Common; Oxhey woods; damp woods about Farnham; Highwood, near Rickmansworth; Mimms Wood; and in most of the woods S.W. of Tring;* heathy wood between Guildford and St. Martha's Chapel; Bisham and Clifden woods, Berks; about Tunbridge Wells; Pryor's Wood, Hatfield Heath; Ball's Wood; Ranmore Common, Denbies.

NEPETA GLECHOMA.—Hedges and waste places, common, 3–5. Hedgerows in the outskirts, everywhere.

NEPETA CATARIA.—Hedges and waste places in a chalky or gravelly soil, not very frequent, 7–9. Croydon, by the roadside near Smitham Bottom,* and on the road to Sanderstead;* Dartford Heath, and road thence to Greenstreet Green; Gravel Hill between Swanscomb and Northfleet; roadside near Northfleet;* about Farnborough; Purfleet.

NUPHAR LUTEA.—Rivers, lakes, and ditches, common, 7–9. Plentiful in the Brent,* Thames,* Roding,* &c.

NYMPHÆA ALBA.—Lakes and still waters, not very common, 7. In the Thames at Ditton (?) *olim*; in the Brent,* scarce; pond on Puttenham Common, abundant;* Felbridge, abundant; Finchley, in a moat,* planted (?) in the Colne at Harefield (?) *olim*; in the Roding (?); pond near Egham; pond at Stow Green, Little Berkhampstead, &c.; pond at Hatfield woodside; Staines; Windsor; Uxbridge Moor.

OBIONE.—*See* ATRIPLEX PORTULACOIDES.

ŒNANTHE FISTULOSA.—Ditches, rivulets, and marshes, frequent, 7–9. Marshy places by the Rifle Butts, Tottenham, plentiful;* Barnes Common;* Putney Heath.*

ŒNANTHE PIMPINELLOIDES.—In similar localities, rare, 6–8. Ditches at Purfleet (?) *olim*, not there now; mistaken for *Œ. Lachenalii?*

ŒNANTHE SILAIFOLIA.—In ponds and marshes, rare, 6. Pools by the road from the Welsh Harp, Hendon, to Woodford House, Kingsbury* (E. de C.); common meadow, Godalming; in the Brent (*olim*), near Greenford; Ham Haw Park, near Weybridge; Eton.

ŒNANTHE LACHENALII.—Salt-marshes, rare, 7–9. Ditches near Purfleet towards Rainham;* Mitcham Common, 1873 (Journal of Bot.); ditch by the road from Staines to Hampton Court; also in cross road to Sunbury (Watson).

ŒNANTHE CROCATA.—Watery places by ditches and river-sides, frequent, 7. river-side between Putney and Kew, &c., in profusion;* Thames side above Moulsey;* Woodford by the Roding; ditch in a meadow by the footpath from the mill, Esher, to West Moulsey.*

ŒNANTHE PHELLANDRIUM.—Ditches and ponds, not frequent, 7–9. By the Colne about Harefield;* in the Roding near Woodford; in the Lea below Chingford old church;* near Romford, Dagenham and Rainham; ponds, Great Warley; ponds, foot of Winter Hill; also at Wargrav and Sonning, in Berks; ponds west side of Ball's Wood; pond between Mangrove Lane and Brickendonbury Park; ponds by the high road between Bell Bar and Milward's Park Wood; banks of canal between Hanwell and Brentford; ditches about Ditton and Chertsey.

ŒNANTHE FLUVIATILIS.—Streams, not very general, 7–9; in the Lea above Chingford Mills;* in the Colne, plentiful,* (between Watford, Harefield, and West Drayton); in the Thames by Chertsey;* in the Lea, near Hertford, beyond scarce.

Œnothera biennis.—Incidentally, about London, and rather frequent, 7–9. About the reservoirs at Walthamstow;* waste places in the suburbs;* railway bank near Weybridge, abundant;* also railway banks, Sutton;* Coulsdon; Lewisham; Warley.

ONOBRYCHIS SATIVA.—Chalky downs, frequent, 6–7. Fields bordering Darent Wood;* about Croydon;* Reigate Hill; fields and banks in many places along the Surrey and Kentish downs, both cultivated and apparently wild.*

ONONIS SPINOSA.—Barren pastures and borders of fields on a clay soil, common, 6–9. (Plenty on Hampstead Heath, E., near the reservoirs;* Barnes Common.*)

ONONIS ARVENSIS.—Pastures, roadsides, and borders of fields, on a chalky soil. Everywhere on the Surrey and Kentish downs;* roadsides, ex. between Croydon and Selsdon,* and between Croydon and Sanderstead,* Box Hill, &c.*

ONONIS RECLINATA.—Sea-shores, rare, 7–11; shore at Northfleet (?) (not seen).

ONOPORDUM ACANTHIUM.—Waste ground and roadsides in a gravelly and chalky soil, not common. Borders of Dartford Heath, rather frequent* (E. de C.); and roadside between the Heath and Greenstreet Green; Gravel Hill, between Swanscomb and Northfleet; Erith (?); Northfleet (?); Lewisham chalk-pit (?) *olim*; Caterham Junction, a plant or two;* Walton Heath, E. of Hertford, by the Ware road; side of the Colne, E. of Colnbrook; Fyfield; Barking; Grays; Tilbury; Southend.

OPHRYS APIFERA.—Both on chalky and on clay soils; on the former, especially in the metropolitan districts, scarce, 6–7. Box Hill;* old quarries W. of Dorking; Purfleet (?) *olim*; about Northfleet and Gravesend (?); old chalk-pit on Morant's Court Hill; chalk-pits, Harefield; Cookham and Bisham hills; Reigate Hill;* Purley Downs; banks bordering Croham Hurst; chalk-pits on the Hog's Back; North Mimms; woods near Watford; chalk-pits, North Hall (or Northaw?) Common, Herts; Tring, meadows between Wiggington and London road; chalk-pit, Gerard's Cross; near Albury, Shiere, Chipstead, and about Ranmore Common, *olim.*

OPHRYS ARANIFERA.—Chalky pastures, rare, 4–5. More frequent in Kent than in Surrey; about Coulsdon, Dartford, Northfleet, and Gravesend; Greenhithe chalk-pits; Mickleham (?); Morant's Court Hill; hills E. of Merstham; Buckland Hill; Norbury Park.

OPHRYS MUSCIFERA.—Chalky banks, more frequent in Surrey, scarce, 5–7. Coulsdon; bank bordering Farthing Downs;* between Dorking and Ranmore Common, and in the chalk-pits; about Northfleet and Gravesend (?) *olim*; Sevenoaks (?downs near); Norbury Park; bordering banks, Croham Hurst, Purley Downs, Epsom Downs; Box Hill; chalk-pits on the Hog's Back; near Tring (by the park); chalk banks in old pits, Reigate Hill;* in two copses near Harefield, on the road to Uxbridge; Cookham, and in Bisham Wood, Berks.; Buckland Hill, E. end; about Albury, Shiere, and Chipstead (*olim*); Morant's Court Hill (old pit).

ORCHIS HIRCINA.—Borders of woods, &c. In the Kentish and Surrey chalk districts, very rare, perhaps extinct; near Knockholt; ? also between Farningham and Shoreham; ? borders of Darent Wood; ? and road to Greenstreet Green; ? at Trulling Down (*olim*); roadside between Crayford and Dartford (*olim*); Box Hill (*olim*); near Puddledock and Stanhill, in Wilmington parish (*olim*).

ORCHIS CONOPSEA.—*See* GYMNADENIA.

ORCHIS PYRAMIDALIS.—Banks and downs in the Surrey, Kentish, and Herts. chalk districts, abundant, 6–8. Greenhithe chalk-pits;* Northfleet and Gravesend; Harefield; Purfleet; Tring Downs;* and Aldbury; Box Hill, and the Betchworth hills;* Dorking chalk-pit;* hills W. of Dorking;* banks bordering Croham Hurst; Mickleham Downs; Reigate hills;* hills E. of Merstham;* Shoreham;* hills E.* and W. of Wrotham;* on the Hog's Back;* chalky banks between Down and Cudham; Purfleet; Cobham Park, bordering banks towards Cuxton.

ORCHIS USTULATA.—In similar situations, but far less frequent, 5–6. Box Hill; Buckland Hill;* S. side of the Hog's Back; Tring and Aldbury; between Knockholt and Wrotham Downs: Mickleham; Harefield; between Colley and Buckland Hill.

ORCHIS INCARNATA.—Rare, local; near Ongar; Hatfield.

ORCHIS MILITARIS.—Borders of woods in the Kentish, Surrey, and Berkshire downs, rare, 5. Box Hill (*olim*); about Dartford, Northfleet, and Gravesend (?) *olim*; near Rochester; woods about Cobham Park, near Cuxton (var. *fusca*); also between Knockholt and Wrotham; pit near the mill, Harefield; Buckland Hill; Tring, banks W. and Aldbury downs; near Reading; Bisham Wood, Berks.

ORCHIS PURPUREA.—*See* ORCHIS MILITARIS, var. *fusca* (the Kentish form, and now not seen in Surrey, except possibly on Buckland Hill).

ORCHIS MORIO. Meadows and pastures, frequent, 5–6. Below Box Hill and the Betchworth hills; field left of road between Lee and Eltham (?) *olim*; meadows opposite Swan Inn, Hendon; meadows of Down Farm and Hightrees Farm, Reigate; between Ash railway station and the Hog's Back; meadows at Pinner* and Ruislip,* abundantly; meadows between the mill, Esher, and W. Moulsey*; Chigwell; pastures near Hertford; N. Mimms; St. Albans; Great Berkhampstead; Essendon.

ORCHIS MASCULA.—Woods and pastures, frequent, 4–6. Box Hill; between Lee and Eltham, with the preceding (?) *olim*; Epping Forest;* Mickleham; N. Mimms; Harefield; Mill Hill; copse near Pinner Wood; plentiful in the meadows between the mill, Esher, and W. Moulsey;* between Down and Cudham; Ruislip;* Pinner;* Epping Forest; copses, Wray Common.

ORCHIS LATIFOLIA.—Marshes and moist meadows, frequent, 6–7. Wimbledon Common, in the further ravine;* Ruislip Moor,* abundant;

banks adjoining Merstham pools; meadows left of the Dorking road between Reigate Hill and Buckland; Harefield, and Rickmansworth, in the marshy meadows, plentiful; Hatfield Forest; Fyfield; Boxmoor.

ORCHIS MACULATA.—Heaths, pastures, and boggy places, frequent, 5–7. Epping Forest;* Keston Common;* Reigate Heath (W. end);* Box Hill; Stanmore Heath; Scratch Wood; Warley Common.

Ornithogalum umbellatum.—Meadows and pastures, rare, 5–6. About Croydon (?) *olim*; Shirley Common (?) *olim*; in a waste field near Charlton Church (?) *olim*; Wimbledon Park; foot of Winter Hill, Berks.; meadows about Harefield; Wellington College; Streatham (in the closes); Teddington, on a small island in the Thames; near Hertford Union workhouse, in meadows by the footpath to Chadwell; in Ware Park, by footpath from Hertford to Ware; near Purley Oaks; Stoke Park, near the churchyard (Surrey); meadows about Reigate.

ORNITHOGALUM PYRENAICUM.—Pastures, rare, 6–7. Land's End near Ripley, (Guildford neighbourhood); near Godalming; on Strawberry Hill (?) *olim*.

Ornithogalum nutans.—Fields and orchards; rare, 4–5. Wimbledon Park; side of the Ravensbourne at Lewisham, (?) *olim*; bank right from Linkfield Street to the Wray Common, Reigate, *olim*.

ORNITHOPUS PERPUSILLUS.—Dry sandy and gravelly places, common, 5–7. Hampstead Heath, abundant.*

ORIGANUM VULGARE.—Dry hilly places, especially in the chalk districts; common, 7–9. Everywhere on the chalk: Purfleet;* lanes between Carshalton and Banstead;* also between Croydon, Selsdon, and Sanderstead, &c.;* Greenhithe;* Tring;* Harefield;* on the downs from the Hog's Back to Cuxton;* about Hertford;* Essex; lanes about Hatfield Broad Oak, as far as the Rodings and Fyfield;* Dartford;* &c.

OROBANCHE MAJOR.—Parasitic on roots of leguminous plants; not unfrequent, 5–7. Box Hill and neighbourhood; Harefield, in a lane near the village; Epping Forest; Wimbledon Common (?); Warley Common; woods adjoining Felbridge pools; furze fields, N. Mimms, Herts; furze field, Colney Heath.

OROBANCHE MINOR.—Parasitic on clover, more frequent than the preceding species, 6–10. Near Croydon;* about Coulsdon, Leatherhead;* Brockham;* Betchworth; Dorking;* woods adjoining Felbridge pools, foot of the Hog's Back, near Puttenham;* mostly in clover fields on the chalk in Kent and Surrey.

OROBANCHE ELATIOR.—On *Centaurea Scabiosa*, and in clover fields and bushy places on the chalk, 6–8. Fields, Carshalton, Sutton, and Banstead downs; about Coulsdon; Box Hill and Betchworth; between Hertfold and Ware.

OROBANCHE CÆRULEA.—Rare, 6–8. In clover fields near Cookham, Berks; Hoddesdon, in the open field W. by road to Hertford.

OROBANCHE RAPUM.—*See* O. MAJOR.

OROBUS TUBEROSUS.—Thickets in hilly places, frequent, 5–7. Wimbledon Common;* Croham Hurst;* Epping Forest;* Hadley Wood, near Barnet;* Chislehurst woods;* Highgate Wood;* Keston Common;* Shooter's Hill;* woods about Broxbourne, Bayford, &c.*

OXALIS ACETOSELLA.—Shady woods; frequent. Plenty in Epping Forest.*

PANICUM.—*See* ECHINOCHLOA.

PAPAVER RHŒAS.—Cornfields and waste places; common, 6–10. Between Hammersmith and Barnes Common;* cornfields by the Thames.*

PAPAVER DUBIUM.—Cornfields and waste places; also on walls; not unfrequent, 5–7. About Croydon; Mortlake on a wall by the Thames;* about Willesden Junction railway premises;* Pinner;* between Dartford and N. slope of Darent Wood in plenty;* near Ewell; Box Hill; Dorking; and many other places above and below the downs.*

PAPAVER ARGEMONE.—Cornfields in the chalk districts; rare, 5–7. Cornfield near Greenhithe, towards Darent Wood;* Box Hill; Dartford and Crayford, between these places; on gravelly soil N.W. of Hertford; between Cobham and Cuxton; Essendon; Cheshunt; St. Albans; Harefield.

PAPAVER HYBRIDUM.—Chalky cornfields; rare, 5–7. Cornfields between the Merrow Downs and Guildford* (on the slope); also on the Hog's Back;* in a cornfield between Dartford and Darent Wood;* between Dartford and Northfleet; cornfields, Harefield; between Cobham and Cuxton.

Papaver somniferum.—Cultivated fields, incidentally. Between Greenhithe and Darent wood;* Dartford, in a field near towards Greenstreet Green;* about Coulsdon; between Cobham and Cuxton (in a field sloping up to the park);* near Headley, coming over the downs from the Dorking road;* Great Warley.

PARIETARIA DIFFUSA.—Old walls (and in waste places occasionally); frequent, 6–9. Fulham by the church wall;* Mortlake in quantity, wall by the river side;* wall, Greenhithe near the station;* walls, St. Mary Cray; Chigwell; Highgate; &c.

PARIETARIA OFFICINALIS.—*See* P. DIFFUSA.

PARIS QUADRIFOLIA.—Wet shady woods, rare, 5–6. Petz Wood, Chislehurst (enclosed); Longwood (also enclosed); Old Park Wood, Harefield; Nutwood, in Gatton Park, Surrey; Epping Forest; Fyfield; woods, Aldbury; Bayford Wood; copse between Reigate Hill and Wray Common; Ball's Wood; Easney Park Wood; copse near Pinner Wood; and copses on the hills E. of Merstham.

PARNASSIA PALUSTRIS.—Boggy meadows, rare, 8–9. Between High and Chipping Ongar; and in a marshy meadow at Chipping Ongar; Harefield, near the mill (?) *olim*; common moor, Rickmansworth; Boxmoor; in a boggy field near Cashiobury Park; Tring reservoirs; Lea valley, near Hatfield;* and Hertford; meadows between Great Berkhampstead and Bourne End, Watford.

PEDICULARIS PALUSTRIS.—Bogs and wet marshy pastures; frequent, 5–9. (Hampstead Heath;* Putney Heath and Wimbledon Common;* marshy places by the canal, Woking, &c., in abundance;* Keston Common.*)

PEDICULARIS SYLVATICA.—Moist heaths and pastures; common, 4–7. (Hampstead Heath; Putney Heath.*)

PEPLIS PORTULA.—Watery places; frequent, 7–8. (Hampstead and Putney Heaths;* Wimbledon Common.*)

PETASITES VULGARIS.—Watery meadows and by water sides; frequent, 3–5. By the Thames, other side of Hammersmith Bridge;* between Mortlake and Kew, in osier holts* (if these have not all been converted into garden land); Chingford, in the mill-race.*

PETROSELINUM SEGETUM.—Moist fields in chalky soils; not common, 8–9. Abundant in a field S. of Cobham Park* (E. de C.); banks in the Thames marsh-lands in several places; Purfleet;* Grays; Tilbury;* Erith;* Northfleet and Gravesend; between Greenwich and Woolwich; borders of Epping Forest, about Chingford (?) *olim*; between Esher and W. Moulsey; borders of fields near Farnham; by the roadside between Hertford and Hertingfordbury; cornfields on the Hog's Back; about Coulsdon; near the river at Woodbridge.

Petroselinum sativum.—Incidentally near gardens and cottages, 6–8. Northfleet chalk-pits in profusion, well established* (E. de C.).

PEUCEDANUM PALUSTRE.—Rare, local; in boggy places, 7–8. Border of upper forest, Epping, Epping side of Roydon in a broad ditch (Flora of Essex). *Œnanthe Lachenalii* or *Silaus pratensis* mistaken for it (Cybele Brit.).[1]

PHALARIS ARUNDINACEA.—*See* DIGRAPHIS.

Phalaris canariensis.—Incidentally on rubbish and waste places about habitations, 7. May often be found on dunghills, &c., in the suburbs.*

PHLEUM PRATENSE.—Meadows, common, 6–10. Pastures everywhere in the suburbs.*

PHLEUM ARENARIUM.—Sandy sea-shores, 5–6. Southend.

PHLEUM BŒHMERI.—Rare. Near Hertford Union workhouse, on a steep gravelly bank by the road to Stanstead; gravel pit between Holwell and Hatfield (T. B. Blow).

PHRAGMITES COMMUNIS.—Ponds and ditches, common, 7–8. Abundant in ditches by the Thames below Greenwich, on both sides of the river for miles.*

PHYTEUMA ORBICULARE.—Chalky downs in Surrey and Kent; rare, 7–8. Purley Downs;* roadside between Croydon and Sanderstead, near the downs;* old chalk-pits, Dorking;* chalk-pits, &c., on the Hog's Back; about Coulsdon, Leatherhead (?); Mickleham; Box Hill; Shiere; Albury.

PICRIS HIERACIOIDES.—Roadsides and borders of fields in the chalk districts (and on gravel drift over chalk); frequent, 6–10. Everywhere in the above localities along the chalk range from Cuxton to Farnham;* Purfleet;* Greenhithe;* Dartford;* Croydon;* between Carshalton and the Banstead Downs;* Essex, between Hatfield Broad Oak and Fyfield, but not so frequent; Harefield.*

PIMPINELLA SAXIFRAGA.—Dry pastures, common, 7–9. (Banks of the Paddington canal; plentiful towards Willesden.)

PIMPINELLA MAGNA.—Shady places in the chalk districts, not frequent, 7–8. Reigate Hill;* lanes below the hills W. of Reigate Hill;* Coulsdon; about Guildford, plentiful; Greenhithe; Westerham;

[1] *Peucedanum officinale* plentiful at Faversham, in the creek there.

Sevenoaks; hills E. of Merstham; Dorking; about Hertford; near S.W. corner of Verulam, St. Albans; Ongar; Brentwood; Godalming; between Godstone and Longfield; near Hertford and to the south of it; about Ball's Wood; footpath to Bayford; copse, Tom's Hill, Aldbury; lane leading to Woodcock Hill, Great Berkhampstead; hedge by Tring reservoir.

PINGUICULA VULGARIS.—Bogs and moist heaths, rare so far south, 5–6. Bog in the Petz wood (enclosed), Chislehurst (?) *olim.*

PINUS SYLVESTRIS.—Woods and plantations, frequent, 5–6. Abundant about Esher, Weybridge, &c.;* but planted (?); a clump on Hampstead Heath;* &c.

PLANTAGO MAJOR.—Pastures and roadsides, common, 6–8. Everywhere in the suburbs.*

PLANTAGO MEDIA.—Pastures on chalky soils, common, 6–10. Plentiful on Farthing Downs, &c.; everywhere on the chalk range.*

PLANTAGO LANCEOLATA.—Pastures and roadsides, common, 6–7. Everywhere in the suburbs.*

PLANTAGO MARITIMA.—Grassy pastures by the Thames below Woolwich, 6–9. Abundant both sides of the river.*

PLANTAGO LACUSTRIS.—Gravelly wastes and sandy commons, common, 6–7 (Plentiful on Hampstead Heath;* Barnes Common,* &c.)

PLANTAGO CORONOPUS.—*See* P. LACUSTRIS.

POA ANNUA.—Pastures and roadsides, common, 4–9. Everywhere in the suburbs.*

POA BULBOSA.—Sandy seashores, local, 4–5. Southend?

POA NEMORALIS.—Woods and thickets, frequent, 6–7. Epping Forest;* thickets, Claygate;* lanes near Reigate.

POA COMPRESSA.—Walls and dry waste ground, frequent, 6–7. Suburbs in many places,* waste ground near West-end (Hampstead) railway station,* near Moulsey.*

POA PRATENSIS.—Meadows and pastures, frequent, 6–7. Pastures about London, plentiful.*

POA TRIVIALIS.—Meadows and pastures, common, 6–7. Pastures about London everywhere.*

POLEMONIUM CÆRULEUM.—Banks and bushy places, very rare, 6–7. Of doubtful occurrence in the metropolitan districts, except incidentally as a garden waif, at Windsor; between Reading and Speenham land.

POLYGALA CALCAREA.—Surrey heaths, rare, 5–9. Box Hill, slope of a valley right of the lane from Mickleham to Headley.

POLYGALA VULGARIS.—Dry, hilly pastures, frequent, 5–9. Epping Forest;* Banstead Downs;* Shirley Common;* Keston and Hayes commons;* Hampstead Heath, sparingly.*

POLYGONATUM MULTIFLORUM.—Woods, rare. Epping Forest, between Epping and Theydon; and in Epping Mill Copse; woods, Tring; Finchhampstead woods, Berks, plentiful; river Cary, in bordering woods.

Polygonum Fagopyrum.—Dunghills, and about cultivated land, 7–8. Frequent on roadsides, and borders of fields in the neighbourhood of Ascot and Chobham, where the grain is raised.*

POLYGONUM CONVOLVULUS.—Cornfields, frequent, 7–9. In almost every cornfield;* also in market gardens.*

POLYGONUM DUMETORUM.—Thickets, rare, 8–9. Cornfields near Farnham, beyond Sir G. Barlow's garden; wood near Mickleham? (in 1835); hedge in Shiere parish; road from Woking Common stat. to Guildford; Witley Lagg.

POLYGONUM AVICULARE.—Waysides, and waste places, common, 5–9. Everywhere in the suburbs.*

POLYGONUM HYDROPIPER.—Borders of pools and in ditches, common, 8–9. (Plentiful on Hampstead Heath* (in holes); Harrow Road, &c.*)

POLYGONUM MINUS.—Gravelly and watery commons, not uncommon, 8–9. Ditton Marsh, by the roadside;* Colney Heath;* grassy commons near the village, Pirbright Heath;* ditto, near Worplesdon;* Wimbledon Common; borders of Frensham pond; Epping Forest; Great Warley.

POLYGONUM MITE.—In similar situations, rare, 8.

POLYGONUM PERSICARIA.—Moist ground and waste places, frequent, 7–10. Everywhere in the suburbs, and often in gardens.*

POLYGONUM LAPATHIFOLIUM.—Fields and dunghills, frequent, 7–9. In the suburbs, everywhere, and often on dunghills about farms.*

POLYGONUM AMPHIBIUM.—Margins of ponds, ditches, and damp grounds, frequent, 7–8. Plentiful on the banks of the reservoir at the Welsh Harp, Hendon, and by the Brent.*

POLYGONUM BISTORTA.—Moist meadows, not common, 6–9. In the meadow below Bishop's Wood (*olim*, now rather scarce); river-side, Uxbridge; and about Rickmansworth; Wotton meadows; Shiere; Blackwater meadows.

POLYPOGON MONSPELIENSIS.—Moist pastures, and borders of ditches by the Thames, very rare, 6–8. By the great ditch near Purfleet (?) *olim*; a mile and a half from Tilbury towards Grays, in ditches, opposite W. end of Northfleet; Greenhithe (?); near the Butts on Plumstead Common; Canvey Island, between the chapel and the river, and near the World's End.

POLYPOGON LITTORALIS.—In similar situations, and equally rare, 7. Canvey Island, between the chapel and the river* (E. de C.); (yes, *littoralis*, H. C. Watson); near the powder magazine, Woolwich (half-way to Erith), *olim* (?); Southend.

POPULUS ALBA.—Moist woods, rare, 3–4. Epping Forest;* Pinner Wood;* about London (but planted?);* between Southend and Leigh.

POPULUS CANESCENS.—Wet, turfy meadows and dry heaths, frequent, 3–4. Plantations about London, not uncommon;* by the Thames, near Richmond;* roadside near Colney Hatch;* pasture through which a footpath runs up to Hampstead from the Finchley road;* a ♀ tree on Barnes Common;* near Ongar.

POPULUS TREMULA.—Moist woods, not common, 3–4. Epping Forest, near High Beech; Hertford Heath, a tree or two,* may be more thereabouts, especially in Ball's Wood; by the Cran, near Hounslow; Bishop's Wood, Highgate; Harrow Weald; Warley Common;* on the Holmwood.

Populus nigra.—Watery places, and by river-sides, not uncommon; but probably planted, 4–5. Some fine ♀ trees by the river, between

Hammersmith and Putney, Surrey side; and some equally fine ♂ trees in the pastures adjoining Barnes Common.* Var. *fastigiata* occurs more generally.*

POTENTILLA FRAGARIASTRUM.—Woods, bank, and dry pastures, frequent, 3–5. Dry banks in most of the lanes a short way out of London;* Hampstead Heath;* Epping Forest, &c.

POTENTILLA TORMENTILLA.—Moors and heaths, common, 6–8. (Hampstead Heath;* Barnes Common.*)

POTENTILLA REPTANS.—Pastures and waysides, common, 6–9. Everywhere in the outskirts.*

POTENTILLA ANSERINA.—Roadsides and moist meadows, frequent. Roadsides almost everywhere; in the outskirts;* canal banks, &c.*

POTENTILLA ARGENTEA.—Pastures and roadsides, in a gravelly soil, not common, 6–7. Wimbledon Common* (scarce); Moulsey Hurst; lane between Betchworth Park and the mill; near Godalming, and on Barnacle Hill; Stonechalk pit; about Bromley; Epping Forest;* near Harefield; Croham Hurst, and gravelly fields about Croydon; banks of the reservoirs near Barnes;* sandy lanes and fields near Farnham; Shirley; Addington; between Dartford and Greenhithe; slopes near Milden's Wood, Godalming; Gallows Plain, and elsewhere about Hertford;* about Hatfield Park; gravel pit on Cook's Hill, Little Berkhampstead; about the Thames, Teddington, &c.; near Byfleet, road to Cobham; road from church, Cobham to Fairmead; St. Martha's Hill, Guildford.

POTAMOGETON COMPRESSUS.—*See* P. PUSILLUS.

POTAMOGETON NATANS.—Ditches, and slow streams, and stagnant waters, frequent, 6–7. In many of the large ponds about London;* Putney Heath;* pond below Totteridge, coming from Mill Hill, in abundance, &c.*

POTAMOGETON POLYGONIFOLIUS.—Bogs and small streams, frequent, 7. Putney Heath;* in all the bogs on the Surrey heaths;* Bell Bar bog; Ball's Wood; ditches, Harrow Weald Common.

POTAMOGETON PRÆLONGUS.—Rare. Ditch adjoining the Thames at Caversham Bridge, near Reading.

POTAMOGETON RUFESCENS.—Ditches and slow streams, not frequent. In a pond not far from the large one near Totteridge, referred to in *P. natans* (and, perhaps, also in that); in the Colne, on Colney Heath; ditches by the Colne, between Harefield Mill and Rickmansworth; in the Lea, near Chingford.*

POTAMOGETON HETEROPHYLLUS.—Pools and ditches, not frequent (?), 6–7. In the Basingstoke canal, near Woking;* lake in Epping Forest, near Wanstead.

POTAMOGETON LUCENS.—Lakes and streams, frequent, 7. Thames, in many places; Lea Canal,* plentiful (about Tottenham); Greenwich and Woolwich marshes (in the ditches).

POTAMOGETON PERFOLIATUS.—Ditches, rivers, and ponds, frequent, 7. In the Thames, about Weybridge,* Twickenham,* and elsewhere; Paddington Canal; Lea River and canal;* in the Colne.

POTAMOGETON OBTUSIFOLIUS.—Rare. Twickenham; circular pond opposite Kensington Palace.

POTAMOGETON CRISPUS.—Ditches and rivers, frequent, 6–7. Pond by the

Palace, Eltham; Keston Common (in the pond); in the Lea and Canal,* and in the canal at Greenford.

POTAMOGETON DENSUS.—Ditches, frequent, 6–7. Plentiful in the ditch round the Palace grounds, Fulham.*

POTAMOGETON PUSILLUS.—Ditches and still waters, not unfrequent, 6–7. Greenwich and Woolwich marshes, in the ditches there;* river Cray; Thames, about Twickenham; ditches about Staines; Lea Canal;* Tring reservoir; Cutmill ponds. (*P. zosterifolius*, a variety of the preceding.)

POTAMOGETON PECTINATUS.—Rivers and ponds, not unfrequent, 6–7. Canal, Tottenham;* river Cray; pond near the Palace, Eltham; Thames, about Staines, and between Hampton and Kingston Bridge;* Ruislip reservoir; Lea; Roding; Grand Junction Canal; Serpentine.

POTAMOGETON ACUTIFOLIUS.—Rare. N.W. corner of Colney Heath, in the Colne; in a pool at London-Colney, E. side of the bridge.

POTAMOGETON PROTEUS.—*See* P. HETEROPHYLLUS.

POTAMOGETON GRAMINEUS.—*See* P. OBTUSIFOLIUS.

POTERIUM SANGUISORBA.—Dry, chalky pastures, common, 6–8. Everywhere on the chalk; from the Hog's Back to the Medway;* roads leading from Croydon to Selsdon and Sanderstead;* from Dartford to Darent Wood;* from Carshalton to Banstead;* Tring;* Herts, on the chalk.

PRIMULA VULGARIS.—Woods, copses, and hedgebanks, frequent, 4–5. (Coombe Wood, Wimbledon;* copses about Harrow,* &c.)

PRIMULA VERIS.—Meadows and pastures, common, 4–5. (Plentiful about Harrow,* Pinner,* Boreham Wood.*)

PRIMULA ELATIOR.—Damp woods, rare, 4–5. Epping Forest (?) *olim*; about Coulsdon.

PRUNELLA VULGARIS.—Pastures, frequent, 7–8. (Hampstead Heath,* and everywhere about London.*)

PRUNUS SPINOSA.—Woods and hedges, common, 4–5. Some bushes on Hampstead Heath;* everywhere in hedges in the environs; segregate *domestica*, near Croham Hurst;* hedges between Betchworth and Dorking; about Reigate; segregate *insititia*, in Epping Forest; about Warley and Brentwood; about Reigate.

PRUNUS AVIUM.—Woods and hedges, rare, 5. Wood at Fyfield; less uncommon, in Herts; Bentley Priory; Harrow Grove; Harefield; about Claygate and Oxshott; Churlwood; Redstone Hill.

PRUNUS CERASUS.—Woods and coppices, not very frequent, 5. Croham Hurst;* below Shooter's Hill, towards Eltham; woods about Gatton; copses on Reigate Hill,* and hedgerows on Colley Farm, below it; Gatshall Copse, Godalming; hedges, Harlow; E. of Aldbury Nowers; near Miswell, Tring; Harrow Weald Common; about Claygate, and between Hook and Chessington.

PRUNUS PADUS.—Woods and coppices, rare, 5. About Hampstead (?) *olim*; West Ham, between the Abbey and the London Road (?) *olim*; hedges about Shiere and Gomshall; near Epping; N.E. side of Hatfield Park.

PSAMMA ARENARIA.—Seashores, local, 7. Shoebury beach, abundant.

Pulmonaria officinalis.—Rare, incidental, 5–6. Wood between Croydon and Godstone.

PYROLA MINOR.—Woods, rare, 6–7. Beech woods, at Tring; Aldbury

Nowers Wood; near Crooksbury Hill, Farnham, abundant; Old Thorns, Seale, near Hampton Lodge (the same as preceding?); wood near Brook Street, between it and Bowles Green (H. C. Watson); in a small coppice near Sunninghill station; grove of old trees E. side of Stanmore Heath.

PYROLA MEDIA.—In similar situations, but very rare. Tring woods.

PYRUS TORMINALIS.—Woods and hedges; and in plantations about London, not very frequent, 5–6. Nutfield, near Redhill; about Hampstead; woods, Felbridge; between Chertsey and Virginia Water; Epping Forest; Fyfield; Hainault Forest; Bayford Wood; woods behind Brickendonbury; Verulam hills, St. Albans; woods by Pinner Lane.

PYRUS ARIA.—Woods and hedges, and in plantations about London, not unfrequent, 5–6. About Croydon; Dorking; old chalk-pits about Dartford; Marams Court Hill, near Sevenoaks; woods and pits, Harefield; Redland Hill; in thickets, Reigate, and on Buckland Hill; lane between Essendon, West-end, and the Lea; Hatfield Park; Stubbins' Wood, and elsewhere, Tring. Var. *pinnatifida*; wood, three miles from Farnham; Uxbridge Common.

PYRUS PINNATIFIDA.—Rare, in wild state. Darent Wood.

PYRUS AUCUPARIA.—Woods and hedges, rare; in plantations, frequent, 5–6. Epping Forest; Castle Wood; Warley Common; Hampstead and Highgate woods; Harrow; Winchmore Hill Wood; St. George's Hill; Redstone Hill and Wood; Redland Hill, Holmwood; and between Dorking and Leith Hill.

PYRUS MALUS.—Woods and hedges, common, 5–6. (Several bushes on Hampstead Heath.*)

PYRUS COMMUNIS.—Woods and hedges, rare, 4–5. Between Clayton and Long Ditton (footpath through the fields); Coulsdon; Harefield; Box Hill; the Hog's Back; Thames bank, near Erith* (one tree); Grays; Epping Forest; St. Albans; Broxbourne and Wormley woods.

QUERCUS ROBUR.—Woods, hedges and plantations, frequent, 4–5. (Hampstead and Highgate.*)

RADIOLA MILLEGRANA.—Moist, gravelly, and boggy places, not unfrequent, 7–8. Wimbledon Common; Coulsdon; Keston Common; Esher Common;* Epping Forest; Reigate Heath;* Wandsworth Common (?); Colney Heath; Gerard's Cross Common;* Pirbright Heath;* Bagshot Heath.*

RANUNCULUS AQUATILIS, (aggregate). Rivers, lakes, ponds, and ditches, common; many segregates, as follows:—

[RANUNCULUS CIRCINATUS.—In the Roding; ditches, Moor Park, Farnham.

RANUNCULUS PELTATUS (*pseudofluitans*).—In the Lea;* ponds between Wandsworth and Wimbledon. Var. *floribundus* near Sunningdale; Walthamstow; in the Wey at Elstead.

RANUNCULUS TRICHOPHYLLUS.—Ponds, foot of Winter Hill, Berks.

(R. DIVERSIFOLIUS, R. DROUETTII, R. BAUDOTII.—Few records for these segregates; *R. Drouettii*, *R. Baudotii*, ditches, Plumstead.)

RANUNCULUS FLUITANS.—Deep and running water; in the Thames and Lea;* in the Colne* records as a segregate insufficient.

Ranunculus intermedius (*tripartitus*).—Ditton Marsh, near entrance to Hare Lane ;* and between Oxshott and Claremont Park.]

Ranunculus Lenormandi.—Reservoir, Farnham Common; Bagshot Heath; Woking Heath; Broadmoor; Esher.

Ranunculus hederaceus.—Ditches and borders of pools, common. (Putney Heath.*)

Ranunculus sceleratus.—Sides of ditches and pools, frequent, 5–9. (Hampstead Heath ;* by the Lea ;* Wimbledon Common.*)

Ranunculus Flammula.—Marshes and wet places, common, 5–9. (Hampstead Heath ;* Putney Heath.*)

Ranunculus Lingua.—Marshes, sides of ponds, &c., rare, 7–9. Totteridge Green ;* Barnes Common (?) Wimbledon Common (?) *olim*; (none there now); ponds on Uxbridge Moor; Langley pond between Cooper's End and Thurrocks; by the Strand, Cookham, Berks.; pond N.W. corner of Thrift Wood, Stanstead.

Ranunculus auricomus.—Woods and coppices and banks in shady lanes, frequent, 4–5. Willesden ;* Neasdon ;* Harrow ;* Epping Forest ;* Claygate ;* &c.

Ranunculus acris.—Pastures, common, 6–7. Everywhere; (waste ground near the Brondesbury Station, plenty.*)

Ranunculus repens.—Pastures, common, 5–8. Everywhere in grass patches by the roadsides, &c.*

Ranunculus bulbosus.—Meadows and pastures, frequent, 5–6. In every meadow, (Brondesbury, &c.*).

Ranunculus hirsutus.—Meadows and waste ground. Not frequent; about Croydon; near Epping; meadows near Hadleigh, in Canvey Island; by the side of a wood between Croydon and Mitcham; Blackwater Lane, near the Ford.

Ranunculus parviflorus.—Cornfields, rare, 5–8. Mitcham Common; Banstead Downs; near Ewell. About Brockham Hill, in Elder thickets; foot of Box Hill; about Coulsdon; about Chelmsford; Chislehurst; Harefield; dry banks near Farnham; between Waltham and Epping; with the preceding between Croydon and Mitcham; several places in the Lea districts; frequent in Essendon parish, near the church and roadside; near Essendonbury; bank S.E. corner of Colney Heath; cornfields, St. Albans.

Ranunculus arvensis.—Cornfields, not frequent; about Croydon; Box Hill; Eltham; Aldbury ;* cornfields, Warley.

Ranunculus Ficaria.—Pastures, woods, &c., common, 3–5. Everywhere about London, especially on hedgebanks in cool damp places.*

Raphanus Raphanistrum.—Cornfields, frequent, 5–10. In almost every cornfield, more or less abundant.*

Reseda lutea.—Banks and waste places, and fallow fields in the chalk districts, common, 6–8. Everywhere on the chalk from the Hog's Back to the Medway; between Carshalton and Banstead; Epsom; Mickleham and Leatherhead ;* Dartford ;* Greenhithe ;* roadsides from Dartford to Darent Wood; and from Croydon to Selsdon and Sanderstead ;* Tring ;* Harefield ;* Tilbury; Purfleet.*

Reseda Luteola.—Waste places elsewhere, as well as upon the chalk, frequent, and more so on the chalk, 6–8. Abundant in chalk-pits, Harefield ;* about Croydon ;* road from Carshalton to Sutton ;* by the Paddington Canal near Willesden ; railway banks,* &c.

RHAMNUS CATHARTICUS.—Woods, hedges, and thickets, frequent, 5–6. Hedges by the Thames near Moulsey Lock ;* by the Brent at Neasdon ; hedges about Harefield ; hedges between Beaconsfield and Uxbridge ; Iver Heath and Gerard's Cross ;* lanes between Plumstead and W. Wickham ;* Epping Forest ; banks of the Roding ; of the Mole ; not unfrequent on the Downs ;* Hog's Back ;* near Woodmanstone.

RHAMNUS FRANGULA.—Woods and thickets, especially in boggy situations, frequent, 5–6. Putney Heath, near entrance to Roehampton Lane ;* Reigate Heath ;* woods of Boar Hill and Leith Hill ;* bogs about Chislehurst ; in the wood S. of the E. common, below ;* White Heath Wood, near Harefield ; Epping Forest ; wood near Warley Common ; wood near Compton ; forked pond, Witley Common.*

RHINANTHUS CRISTA-GALLI.—Meadows and pastures, frequent, 6–7. Meadows near Wimbledon Common, below the mill ;* meadows at Pinner ;* Thames Ditton ;* Elstree ; Box Hill, below ; about Harrow ; Lea marshes ; Warley and Brentwood.

RHYNCHOSPORA ALBA.—Bogs, chiefly on the heaths in S.W. Surrey, 6–8. Bagshot Heath, abundant in several places ;* Elstead Common ;* borders of Stotbridge pond ;* Whitemoor Common, in profusion ;* Pirbright Heath ;* Chobham Common ;* Farnham Common, by Burnham Beeches ;* Esher Common ; a patch at the foot of Oxshott Hill ;* Cow Moor ; Bisley Common ;* swamp near Cobham, Surrey ; and peat bogs on Farnham Common ; Windsor Park ; Waterdown Forest (bog).

RIBES NIGRUM.—Woods and river-sides, rare, 4–5. Meadow near the Warren pond, Breakspeares, Harefield ; by the Mole, near the mill, Esher ;* by the river Wey ; wet places about Godalming ; Warley Common ; about the old castle grounds, Great Berkhampstead.

RIBES RUBRUM.—Woods and hedges, rare, 4–5. Coppices near Breakspeares, Harefield ; wood between Chislehurst and Bromley ; about Coulsdon ; Leatherhead ; Box Hill ; bridge by the Mole ; Warley Common ; banks of the Roding ; by the Mole at Esher ; copse opposite Roxford ; near rectory, Hatfield ; in How Dell ; canal side, near Bowne End ; Great Berkhampstead.

Ribes Grossularia.—Hedges and thickets, rare, 4–5. Road between Chislehurst and Bromley (?) ; about Coulsdon ; Ongar ; High Laver.

ROSA SPINOSISSIMA.—Heaths, chiefly on sand and chalk, rare, 5–6. Barnes Common ;*[1] Croham Hurst ; Riddlesdown, near Caterham Junction ;* Waddon Marsh ; about Albury, on the Downs.

ROSA TOMENTOSA.—Hedges and thickets, not unfrequent, 6–7. About London (?) *olim ;* Caen Wood (?) ; between Cobham and Cuxton ; Epping Forest ; lane between Upminster and Gt. Warley ; frequent about Hertford ; Hertford Heath ; Brickendon ; Bayford ; Little Berkhampstead ; Cheshunt ; lane from Colney Heath to St. Albans ; woods S.W. of Tring ; near Welwyn (S. B. Blow).

[1] Several plants on a low bank, doubtless planted there originally.

ROSA MOLLISSIMA.—Charlton Wood (?) *olim*'; Roxeth (Mids.); about Claygate; Fflanchford; Reigate woods; hedges between Redstone Hill and Nutfield; near the foot of Leith Hill.

ROSA RUBIGINOSA.—Bushy places, and borders of woods in the chalk districts, not very common, 6–7. Box Hill; Ranmore Common;* banks bordering Croham Hurst; between Cobham and Cuxton; Riddlesdown, near Caterham Junction;* Holmwood Common; Burnham Beeches; Hertford Heath; Cook's Hill, Little Berkhampstead; Wood Lane, Great Berkhampstead; Stanmore Heath.

ROSA MICRANTHA (INODORA).—Smooth-stemmed varieties (so to speak) of the preceding, and growing in similar districts, but of more general occurrence, 6–7. Surrey and Kentish Downs, in bushy places, frequent; hills W. of Dorking chalk-pits;* hills between Dorking and Reigate;* hills E. of Merstham; hills E. of Shoreham;* and beyond Wrotham;* Burnham Beeches; Epping Forest; Morant's Court Hill; about Hertford in hedges; Gatton.

ROSA CANINA.—Hedges and bushy places, common, 5–6. Everywhere in lanes and roadsides about London.*—N.B. Twenty-nine varieties are enumerated in the seventh edition of the Catalogue; to what extent these may severally occur in the metropolitan districts, there are no records to show; a variety with woolly styles grows in Burnham Beeches.* *See* Baker, 'British Roses.'

ROSA STYLOSA.—Thickets and hedges, frequent, 6–7. Lane leading from Child's Hill to Hendon;* Broxbourne;* and woods thereabouts.—N.B. A good species with a distinctive characteristic; though united to the following by some botanists.

ROSA ARVENSIS.—Woods, hedges and thickets, common, 6–7. Plenty in Epping Forest;* &c.

ROSA SYSTYLA.—*See* R. STYLOSA.

RUBUS IDÆUS.—Upland woods, not frequent, 6–7. Box Hill;* about Mickleham; Merstham; Reigate Heath (?); Aldbury Nowers Wood;* Bagshot Heath (N.W. corner);* Hatfield Forest; Warley Common; Boreham Wood; Keston Common.*

RUBUS FRUTICOSUS (aggregate).[1]—Hedges and thickets; common, 7–8. Many segregates; recorded localities as follows:—

[1] The number of segregates exclusive of R. *cæsius*, amounts in the Catalogue of 1874, to forty, besides intermediate forms. There are six species, according to Hooker and Arnott. No doubt they all run into each other, and it is impossible to draw fixed lines of demarkation, so that they may be distinguished by unexceptionable and by well-marked distinctive characteristics.

For all practical purposes the following arrangement, according to the above-named authorities, may be found sufficient:

a. *Rubus suberectus* and *plicatus*: thickets, hedges, and boggy places; low-lying damp situations by rivulets. For localities see note on the segregates.

b. *Rubus fruticosus*: thickets and hedges (the common form, with leaflets downy on the underside) everywhere.*

c. *Rubus corylifolius*: wet heaths and commons (stems smooth and usually of a reddish-brown colour). Surrey heaths*; plentiful by the Basingstoke canal beyond Brookwood; *Chislehurst common; *Warley common; *Merton.

d. *Rubus carpinifolius*: generally in shady woods or borders thereof, and in a sandy soil (stems generally hairy, or at least the branches are so). Shady lane between St. Mary Cray and Chislehurst; *Chislehurst Wood; *Warley Common; near Epping; Wimbledon Common.

[RUBUS SUBERECTUS.—Easney Park Wood; boggy thickets; foot of Leith Hill; and on Reigate Heath; Wimbledon Common.

RUBUS PLICATUS.—Tring Heath (?) *olim*; boggy thickets, Warley Common; hollow S. of Chislehurst East Common. *

RUBUS AFFINIS.—Hagger Lane; Epping Forest.

RUBUS LINDLEIANUS.—Barrack Wood, Warley; Epping Forest; near Woodford, &c. (*R. nitidus*); dry, gravelly soil, Ball's Wood; Dorking; Tring woods; Essendon, &c.

RUBUS RHAMNIFOLIUS.—Wood near Hale End, Chingford; Warley Common; Wimbledon Common; Bagshot; Dorking; Reigate; Harrow Weald Common.

RUBUS DISCOLOR.—*See* R. FRUTICOSUS (aggregate). *Note.*

RUBUS THYRSOIDEUS.—Hedges, Ditton; Easney Park Wood.

RUBUS LEUCOSTACHYS.—About Hertford; Cheshunt; Tring; N. Mimms; Rickmansworth; St. Albans; Harrow Weald Common; Esher; Dorking; Hale End; Epping Forest; and on the chalk range.

RUBUS CARPINIFOLUS.—*See* aggregates, also Milward's Park Wood; Præ Wood, St. Albans; ravine between Boar Hill and Leith Hill; Hale End and Snaresbrook; Epping Forest; Hertford Heath; woods north of Pinner; Hampton; E. Barnes. Leaves not Hornbeam-like.

RUBUS VILLICAULIS.—Willesden Lane; Broxbourne woods; about Claygate; St. Anne's Hill, Chertsey; Harrow Weald Common; ravine between Boar Hill and Leith Hill.

RUBUS MACROPHYLLUS.—Wimbledon Common; Claygate; wet copse near Littleton Bridge, Reigate; near Red Hill railway station. Var. *Schlechtendalii:* E. Barnet; N. Mimms; Præ Wood, St. Albans; woods by Pinner Lane.

RUBUS HYSTRIX.—Woods by Pinner Lane; Præ Wood, St. Albans; Broxbourne woods.

RUBUS ROSACEUS.—Copses near Panshanger and Essenden Glebe woods, Herts.; (? *cordifolius*) Wimbledon Common.

RUBUS RUDIS.—Woods N.W. of Hertford, and wood on the road to Hertingfordbury; Cheshunt; Tring; Wimbledon; near Rickmansworth; Bell Bar woods; woods by Pinner Lane; Præ Wood, St. Albans; Mimms Park Wood (margins of copses in stiff soil).

RUBUS RADULA.—Woods. Pinner Lane; Egham; Præ Wood, and Sand-pit Lane, St. Albans; Thieves Lane, Hertford; woods at Little Berkhampstead and Bayford; Pinner (sandy gravelly soil).

RUBUS FUSCO-ATER (and FUSCUS).—Panshanger Plain; between Bayford Wood and Little Berkhampstead; woods by Pinner Lane; Broxbourne woods; Acton.

RUBUS GUNTHERI.—Heathy woods, rare. Wimbledon Common; near E. Barnet; Pinner; Tring; Præ Wood; Mimms Park Wood.

RUBUS GLANDULOSUS.—*See* aggregate; var. *hirtus*, near Wimbledon.

e. Rubus glandulosus: woods and hedges generally in upland districts (stems both hairy and setaceous). The more usual variety met with in the chalk districts and in woods: *Warley Common; *Coulsdon; *thickets on the Downs; *Hertford Heath; *Wimbledon Common.

f. Rubus rhamnifolius: between *suberectus* and *fruticosus.* See Segregates.

RUBUS CORYLIFOLIUS.—*See* aggregate; also at Bell Bar; Rickmansworth; Milwards Park Wood; Gatton; Norbury Park, Headley Lane; Leyton; Southend, &c.

RUBUS TUBERCULATUS.—Harrow; Willesden.

RUBUS ALTHÆIFOLUS.—Harrow; Pinner.

RUBUS KŒLERI.—Woods S. of Hertford; Hoddesdon; Broxbourne and Wormley woods; Gatton; Norbury Park; Headley Lane; about Walthamstow (woods in a barren soil, especially on wet clay). Var. *pallidus*, Wimbledon Common; Epping Forest.]

N.B. A French botanist (Genevier, 'Brambles of the Loire,') declares that there are "two hundred distinct species of Brambles. Evidently any number may be made, at the option of the individual botanist; say from five to fifty" (Cybele).

RUBUS CÆSIUS.—Hedges and damp places, ditches, &c., common, 6–7. (Lanes about Tottenham;* Hendon.*)

RUMEX CONGLOMERATUS.—Watery places, not uncommon, 6–8. (Riverside between Hammersmith and Kew.*)

RUMEX PALUSTRIS.—Marshy places, rare. Plaistow; Purfleet; Wanstead Park.

RUMEX NEMOROSUS, var. *sanguineus*.—Shady woods and pastures, rare, 7. Bishop's Wood, Hampstead; var. *sanguinea*, Pinner Wood.

RUMEX PULCHER.—Pastures and waysides, not very frequent, 6–8. Roadside between Hampton and Hampton Court (?) *olim*; High Ongar; Wanstead; Purfleet.

RUMEX OBTUSIFOLIUS.—Waste places and waysides, common, 6–7. Everywhere in the environs.* Var. *sylvestris*, Ware.

RUMEX CRISPUS.—Waysides, and near houses, also in pastures, frequent. With the preceding everywhere in the environs.*

RUMEX AQUATICUS.—Moist places, rare, 7–8. Confounded with *R. crispus*, or with the following (Cybele).

RUMEX HYDROLAPATHUM.—Ditches and riversides, not uncommon, 7–8. Barnes Common;* banks of the Thames;* of the Lea Canal and adjoining ditches;* of the Colne;* of the Roding;* &c.

RUMEX MAXIMUS.—A variety of the above, rare. In Herts.

RUMEX MARITIMUS.—Seashores, local. Purfleet.

RUMEX ACETOSA.—Meadows and pastures, common, 5–7. Meadows by the Lea;* everywhere;* few meadows free from it.

RUMEX ACETOSELLA.—Dry pastures, banks, and roadsides, frequent, 5–7. Plentiful on Hampstead Heath;* Barnes Common; &c.*

RUPPIA ROSTELLATA.—Salt-water ditches, local, 8–9. At Southend, E. of the town, ditches on the flats.* In a ditch near the railway station at Cuxton, coming from Cobham.*

RUSCUS ACULEATUS.—Bushy and heathy places, and in woods, not common. 3–4. Claygate, near Ditton; about Coulsdon; Box Hill;* Abbey Wood; Darent Wood;* Epping Forest; Harefield; Warley Common; Farnham Common, beyond the park; Hainault Forest; Holmwood Common; woods about Godalming; Cockshott Hill, Reigate, S.E. of mill; Bayford; Westhumble Lane, Dorking; plentiful. Ashdown Copse, Ranmore Common; Essendon; hedge in Wormley Wood; wood near Northaw.

SAGINA APETALA. Dry gravelly places, and on walls, common, 5–9. (Hampstead, and Hampstead Heath.*)

SAGINA CILIATA.—In similar situations, and on commons, (frequent in Surrey; or else often mistaken for the preceding, of which it may be merely a modification). Tilbury; Harrow Weald Common; about Weybridge; Albury; Witley Common; west of Woking station.

SAGINA PROCUMBENS.—Waste places and dry pastures, common, 5–9. (Hampstead Heath.*)

SAGINA SUBULATA.—Dry, gravelly, and stony pastures, not uncommon, 6–8.

SAGINA NODOSA.—Wet, sandy places, not very frequent, 7–8. By the canal between Harefield and Uxbridge;* by the lake in Hatfield Forest;* Uxbridge Moor; banks of canal near Woking station, and near Pirbright; about Shiere and Albury; Hammer Ponds and on Witley Common; between upper and lower Cutmill ponds; slopes of Box Hill; near Wandsworth Pier, in 1855; near Hoddesdon; Cookham Down.

SAGITTARIA SAGITTIFOLIA.—Ditches and margins of rivers, frequent, 7–9. (Ditches by the Lea Canal, and by the Lea;* in the Roding;* Paddington Canal;* Thames;* Colne;* &c.)

SALICORNIA HERBACEA.—Riversides, by the Thames below Greenwich, local, abundant. At Purfleet;* Canvey Island;* Rochester.*

SALIX PENTANDRA.—Banks of rivers, and in watery places, rare, 5–6. Near the brick kiln, Harefield; in an enclosure on Esher Common, near the farms on the road across it to Oxshott Hill; Coombe Wood, Wimbledon; about Fulham, *olim*; by the Lea near Whitwell; hedge by roadside near King's Langley.

SALIX FRAGILIS.—Banks of rivers, marshy woods, and osier grounds, common, 4–5. Abundantly by the Thames between Putney and Kew;* by the Lea.*

SALIX RUSSELLIANA.—Variety of *S. fragilis*, in similar situations.

SALIX VIRIDIS.—Ambiguity. Variety of *S. fragilis* (?); *vide* 'Botanical Gazette,' iii. p. 60.

SALIX ALBA.—In similar situations, common. Plentiful near Mortlake and Kew;* and in the Lea meadows.*

(*b*, *c*, *cærulea* and *vitellina*, varieties; *c*, in hedges, Hounslow Road, Twickenham; and by the Thames between Twickenham and Richmond; Mangrove Lane, Hertford, in a hedge.)

SALIX UNDULATA.—In similar situations, rare, 4–5. By the Thames about Richmond; by the Lea near Higham Hall.

SALIX TRIANDRA.—In similar situations, not uncommon, 4–6. By the Thames between Putney and Mortlake;* Mole at Brockham;* Thames between Richmond and Hampton Court;* ditches by the Lea;* and by the railroad between Tottenham and Broxbourne.*

(*c*, *amygdalina*: variety of the preceding, and in similar situations. By the Brent between Greenford and Perivale; foot of Box Hill; by the Mole, Woodford.)

SALIX PURPUREA.—Marshes and riversides, not very frequent, 4–5. Thames about Twickenham; by the Cran, near Isleworth;* and at Hospital Bridge;* Wargrave. Var. *Lambertiana*, banks of the Lea, at

Stanstead; Great Berkhampstead Castle moat; Elstree; Rickmansworth; Colne, near Colney Street.

(*b*, *Woolgariana*, by the Thames at Kingston.)

SALIX RUBRA.—In similar situations, rare, 4–5. By the Queen's River, Hampton; Thames opposite Ditton; meadows below the hills on the Mole at Esher; Thames between Kew and Richmond; by the Lea (? var. *Helix*).

(*b*, *Forbyana*, near Brockham; and about Betchworth; by the Lea, near Higham Hill.)

(*c*. *Helix*, ambiguity: var. of the above.)

SALIX VIMINALIS.—In similar situations, common, 4–5. Thames side, abundant;* Roding;* Lea;* &c.

SALIX STIPULARIS, ambiguity.—(Var. of the preceding?) Lea Bridge Road; banks of the Lea under Higham Hill.

SALIX SMITHIANA.—Meadows and osier grounds, rare. Copse at Pinner Park, near the Lea Bridge Road; and in a footpath from Marsh Street to Lea Bridge, and near Higham Hill; Warley.

SALIX ACUMINATA.—In similar situations, rare (?). Localities for this and the preceding not separable ('Cybele,' Comp.).

SALIX CINEREA.—Wet hedgerows, swampy places, river banks, and moist woods, common, 3–4. (Putney Heath, in the hollow towards Richmond Park;* woods at Chislehurst.*)

SALIX AURITA.—Moist woods and thickets, common, 4–5. (Putney Heath; Epping Forest.*)

SALIX CAPREA.—Woods and roadsides, common, 4–5. Highgate Wood;* Epping Forest;* Wimbledon Common.*

SALIX NIGRICANS.—Fens, osier grounds, and sides of streams, rare. By the Mole at the foot of Box Hill (?).

SALIX REPENS.—Heaths and moors, common, 4–5. A variable plant. Hampstead Heath;* Barnes;* Putney and Wimbledon commons.* Variety *fusca*, Stanmore Heath;* and in Epping Forest;* and Harrow Weald Common.*—N.B. "The study of this genus has been made difficult and unattractive, by excessive subdivisions, and consequent uncertainties" ('Cybele Britannica').

SALSOLA KALI.—Seashores, local, 7. Between Leigh and Southend.*

SALVIA VERBENACA.—Dry pastures in a chalky or gravelly soil, not common, 5–8. Between Greenhithe and Northfleet, by the roadside;* roadside near Greenstreet Green;* roadsides near Guildford; between Erith and Plumstead; about Cobham; by the Thames occasionally; by the towing path above Richmond;* Purfleet; Southend.*

SALVIA PRATENSIS.—Dry meadows, &c., very rare; in Kent, formerly in Cobham Park; in enclosed ground near Wrotham (Mr. Hanbury).

SAMBUCUS NIGRA.—Woods and coppices, common, 6. Shady shades in the environs, frequent, copses, &c. Hendon;* Epping Forest;* about Chingford;* in most hedges.*

SAMBUCUS EBULUS.—Waysides, and waste places generally in damp localities, not very frequent, 7–8. Epping Forest; lane between Sewardstone and Waltham Abbey; Uxbridge Moor, abundant; between E. and W. Tilbury; meadow at Breakspeares, Harefield; in a hedge near the Roding, going from Chigwell to Loughton;* about Ewell; at Ratley by the roadside (six miles from Barnet; is this Radlett ?); Upper Lea Valley between Aston and Shephall, plentiful; Gatton Park below the engine pond; in a field left, ascending the hill from Croydon to Beulah Spa.

SAMOLUS VALERANDI.—Marshy and watery places, not uncommon, 6–9. Ditch-sides in the marshes below Woolwich;* Erith;* and Greenhithe; marshy meadows behind N. Cray; Warley Common; watercourses, Hurtmore Common; Plaistow; Southend.

SANGUISORBA OFFICINALIS.—Low moist meadows, on a chalky soil, rare, 6–8. About Croydon; road from Dartford Heath to Greenstreet Green; pastures, Whitechurch and Stanmore (?); Sonning Meadow.

SANICULA EUROPÆA.—Woods and thickets, frequent. Epping Forest abundant;* Hadley Common Wood.*

Saponaria officinalis.—Roadsides, margins of woods, and hedgebanks, especially near cottages, rare, 7–8. Roadside between Luckfield Street and Ffrenches, Reigate; near Down Mill; between Cobham and Feltham; bank in a lane E. of Reigate; about Abinger; Box Hill; beyond Dartford Heath and about Dartford; lane near Uxbridge Church; lane near Shorne;* hedge near Roxford Farm, Hertingfordbury; Stapleford by footpath to Mill End; Theobalds Lane, Cheshunt; S. side of St. Martha's Hill between Chilworth House and farm; wood at Purfleet; about Essendon; near Shiere.

SAROTHAMNUS SCOPARIUS.—Dry hills, and bushy places, common, 5–6. (Hampstead Heath;* Barnes Common;* railway banks between Kew and Acton; abundant.*)

SAXIFRAGA GRANULATA.—Hedgebanks, pastures, and meadows, in a gravelly soil, not uncommon, 5–6. About Croydon; roadside to Addington;* about a pond further on, foot of the common;* St. Catherine's Hill, Guildford; hedgebank near W. Wickham; wood near Dartford, and by the roadside near Greenstreet Green;* meadows at Harefield; Rickmansworth, plentiful; and near Watford; in the park, Wimbledon (?) *olim*; about Swanscombe; Chislehurst; Shiere, abundantly; banks near Farley House; roadside near Ringley Oak Gate, Reigate; lane W. of Reigate Park; sandy lanes Seal, near Farnham; Southend (plenty); Waddon Marsh; Mead Lane, Hertford, and meadows between Hertford and Ware; road from Hatfield Union to Stanstead, particularly between West End and Hatfield Park.

SAXIFRAGA TRIDACTYLITES.—Walls, and dry barren ground, common, 4–7. Frequent on walls in the suburbs; (road from Hammersmith to Fulham;* Brentford towards Hounslow;* &c.).

SCABIOSA SUCCISA.—Moors, damp meadows, and pastures, common, 7–10. (Plentiful on Hampstead Heath, about the bog.*)

SCABIOSA COLUMBARIA.—Pastures and waste places, on a chalky soil, frequent, 7–8. Of general occurrence on banks, &c., from the Hog's Back to Cuxton;* slopes about Caterham Junction;* Banstead Downs;* road from Carshalton to Banstead, in a pasture left, in extraordinary abundance;* banks bordering Tring woods;* Dartford;* Greenhithe and Purfleet; about Hertford.

Scabiosa arvensis.—Pastures and cornfields, more general in the chalk districts, frequent, 6–8. Frequent on banks and borders of cornfields from the Hog's Back to Cuxton;* road from Carshalton to Banstead;* banks of the Thames above Teddington Lock;* between Cobham and Cuxton;* roadsides among the Essex cornfields;* between Croydon and Sanderstead;* about Dartford, Greenhithe, and Northfleet.*

Scandix Pecten-Veneris.—Cornfields, chiefly on the chalk, not uncommon, 6–9. About Croydon;* Carshalton;* Box Hill; Coulsdon;* Greenhithe;* Dartford; Northfleet; Gravesend * (towards Cobham); frequent in Herts.

Schœnus nigricans.—Bogs on moors, rare, 6–7. Bagshot Heath, left of the road from Chobham, abundant;* nowhere else nearer than Barkway Moor, on the borders of Cambridgeshire; fens of Tilbury Fort (*olim*, none there now ?), also in a bog, *olim*, two miles beyond Merstham near Redhill station (Cooper)?

Scilla nutans.—Woods and coppices, common, 4–6. (Epping Forest between Woodford and Walthamstow, abundant;* in small quantity on Hampstead Heath.*)

Scilla autumnalis.—Gravelly pastures by the Thames, local, 8–9. Moulsey Hurst, opposite Hampton Church, in profusion;* Kew Green; meadows above Richmond; near Ham and Ditton; Blackheath, scarce now probably, the heath being so much trodden upon; Shorne Warren, abundantly; between Lee and Eltham (? *olim*).

Scirpus lacustris.—Margins of ponds and rivers, common, 7–8. (Thames shore opposite Hammersmith.*)

Scirpus carinatus.—Thames shore between Putney and Hammersmith, rare, local, 7–8.—N.B. A mere variety of the above, in the opinion of Mr. Bentham; perhaps a hybrid between the common and the maritime species, which at this point come into contact with each other.

Scirpus acicularis.—Sandy places on the borders of ponds, not very common, perhaps often overlooked, 7–8. Borders of a pond on West-end Common, Esher;* ponds on Puttenham Common;* margins of a pond on Milford Green, near Chobham, abundant;* pond near Wanstead in Epping Forest; Bullmarsh Heath, Berks; Elstree reservoir; Earlswood Common.

Scirpus palustris.—Wet marshy places, and sides of ponds and ditches, common, 6–7. (In a pond below Hampstead Heath, right of lane leading to Fortune Green;* ponds on Putney Heath.*)

Scirpus multicaulis.—In similar situations; less frequent; perhaps often passed over for the preceding species, 7. Reigate Heath;* Earlswood Common; Esher Common, between Claremont Park and Oxshott Hill;* Bell Bar bog, Herts; boggy places by the canal, Woking;* Pirbright Common;* Bagshot Heath, by the road to Chertsey, direct;* (in a pond right) Epping Forest, lake at Wanstead; Keston Common; Totteridge; bog near Hatfield; Bell Bar bog.

Scirpus cæspitosus.—Moors and wet heaths, frequent, 6–7. (Wimbledon Common;* Esher Common.*)

Scirpus pauciflorus.—Boggy places; rare, 7–8. Cheshunt (N. B. G.); bog in Epping Forest, between Wanstead and Walthamstow (?); Bell Bar bog; Colney Heath.

SCIRPUS SETACEUS.—Marshes, &c., common. (Hampstead Heath.*)

SCIRPUS FLUITANS.—Ditches and shallow ponds, common, 7–8. (Putney Heath;* Esher Common.*)

SCIRPUS SYLVATICUS.—Moist woods and river-banks, frequent, 7. Plentiful in the Brent below Totteridge;* also at Neasdon;* in the Roding.*

SCIRPUS MARITIMUS.—Muddy banks of the Thames, within full tidal influence, common, 7. Abundant below Greenwich, on both sides of the river;* also extending above London, to near Hammersmith Bridge.*

SCIRPUS TRIQUETER.—In similar situations, very rare; perhaps extinct (?). About Putney, formerly; marshes between Greenwich and Woolwich (?).

SCLERANTHUS ANNUUS.—Cornfields, frequent, 7–8. Pinner; Harefield; Harrow Weald Common; between Teddington station and Bushy Park; Barnes Common;* Wimbledon Common; N. Mimms, Herts; more frequent in cornfields on the downs; cornfields near the Merrow Downs, abundantly;* cornfields beyond Weybridge, towards Chobham, plentiful.*

SCLERANTHUS PERENNIS.—Dry sandy fields; rare, 6–8. Banstead Downs; between Compton and Guildford.

SCLEROCHLOA DISTANS.—Salt-marshes; rare, 7. Rainham; Purfleet;* Tilbury.

SCLEROCHLOA MARITIMA.—Salt-marshes, common, 7. Abundant about Northfleet;* Erith,* &c.; Cuxton.*

SCLEROCHLOA PROCUMBENS.—Salt-marshes, not common, 6–7. Purfleet; Tilbury.

SCLEROCHLOA RIGIDA.—Walls and barren stony places in the chalk districts, 6–7. Banks about Greenhithe;* Northfleet* (also on walls here);* in a field bordering Riddlesdown, near the chalk-pit in great abundance;* of frequent occurrence on banks in the chalk range between Dorking and Cuxton;* Purfleet.

SCLEROCHLOA LOLIACEA.—Sandy sea-shores, 6–7. Marshes near Walthamstow (?); about Woolwich (?); Southend; perhaps, but not recorded.

SCUTELLARIA GALERICULATA.—Banks of rivers and canals; common, 7–8. (Plentiful by the Lea * and canal;* by the Thames.*)

SCUTELLARIA MINOR.—Moist places on heaths, frequent, 7–10. Hampstead Heath, by the pond; Putney Heath, near entrance to Roehampton Lane;* bogs about Chislehurst;* Keston Common;* Gerard's Cross Common, abundant;* also by the roadside here, coming from Iver Heath; bog on Farnham Common, near Burnham Beeches;* bogs, foot of Leith Hill;* boggy places on the Surrey commons, Pirbright;* Esher;* Bagshot Heath;* banks of the Basingstoke Canal; Colney Heath; Bell Bar bog; Harrow Weald Common.

SCROPHULARIA AQUATICA.—*See* S. BALBISII.

SCROPHULARIA BALBISII.—Sides of rivers and wet places; frequent, 6–9. (Plentiful by the Lea and canal, and in the ditches which border the canal.*)

SCROPHULARIA NODOSA.—Woods and damp places, 6–8. Common; (frequent in ditches about Hendon;* Neasdon;* Kingsbury;* Willesden;* &c.)

Scrophularia vernalis.—Waste places and roadsides; rare, 4–6. Between Merton and Mitcham (?) *olim*; Chislehurst; hedges in N. Essex, near Deoden; also to the E. and S. of Saffron Walden; in Hatfield Park, close to the N. front of the house; hedge by the towing-path, near gasworks, Hertford?

SEDUM TELEPHIUM.—Bushy places and borders of fields; also in woods, not common, 7–8. Sandy lane near Frensham Common;* woods S.W. of Tring;* Coombe Wood, Wimbledon Common; near Luxboro' House, Chigwell; Caen Wood, Hampstead; Croham Hurst, Croydon; Charlton Wood; Chigwell; Epping; Frith Hill, Godalming; Cockshott Hill, near Reigate, especially in furze field S.E. of mill; about Albury; bank opposite High Rocks; hedges about St. Albans; about Hertford, Bayford, Essendon, and Hatfield; also Watford and Rickmansworth, Harefield; N. Mimms wood; about Guildford.

SEDUM ALBUM.—Walls, rocks and roofs of houses, not unfrequent; walls of West Ham Abbey (?); old walls about Plaistow and Barking (?); Twickenham; between Brentford and Isleworth;* (wall, Sion House grounds).

Sedum reflexum.—Incidentally near habitations; not frequent, 6–7. Roadside, Pinner, near a cottage;* Hertford, on the castle, plentiful; also in Fore Street and Castle Street; Hatfield Woodside; Hoddesdon; Great Berkhampstead, in Water Lane; Rickmansworth.

SEDUM ACRE.—Walls and sandy ground, common, 6–7. Frequent on walls in the suburbs; banks of the Thames, above Moulsey;* Banstead Downs, near Sutton.*

Sedum dasyphyllum.—Incidentally on walls about London; rare, 6–7.

Sedum sexangulare.—Incidentally on old walls; very rare, 7. Greenwich Park walls; old walls at Northfleet.

Sempervivum tectorum.—Cottage roofs and plants, occasionally; planted, 7.

SENEBIERA CORONOPUS.—Waste places and roadsides, common, 6–8. Frequent in the suburban districts, Edgware Road, *e.g.* beyond Brondesbury station.*

SENEBIERA DIDYMA.—Roadsides and waste places; not frequent, 7–9. Lane leading from the Devil's Punchbowl, Dorking, to the Holmwood, abundantly; in a field near Epping; near Chobham; Parson's Green; Kew Green; lane between Southgate and Colney Hatch.

SENECIO VULGARIS.—Roadsides and waste places, gardens, common, 1–12. Everywhere.*

SENECIO SYLVATICUS.—Dry upland soils, banks and gravelly pastures; frequent, 7–9. (Hampstead Heath;* Barnes Common.)

SENECIO VISCOSUS.—Waste ground on chalky or gravelly soil, not common, 7–8. Streatham Common (?) *olim*; Bexley Heath; old chalk-pits about Dartford; near Kensington railway station, incidentally lately; Symes farm, near Epping; glebe land, at Lee.

SENECIO ERUCIFOLIUS.—Hedges and roadsides, and in waste fields in a chalky or gravelly soil; frequent, 7–8. Banstead Downs;* Box Hill.*

SENECIO TENUIFOLIUS.—*See* S. ERUCIFOLIUS.

SENECIO JACOBÆA.—Waysides and on coarse pastures, common, 7–9. (Hampstead Heath; abundant.*)

SENECIO AQUATICUS.—Wet places and by the sides of rivers and ditches; frequent, 7–8. Ditches in the marshes below Woolwich.*

SENECIO CAMPESTRIS.—High chalky downs; rare, 6. Downs by Aldbury Nowers Wood;* S. side of the Hog's Back, near New Inn.

SERRATULA TINCTORIA.—Thickets and roadsides, in woodland places; not uncommon, 8. Putney Heath;* Hampstead Heath* (among the fern) Bagshot Heath, borders;* roadside from Weybridge to Chobham Common;* Burnham Beeches;* Ball's Wood;* Epping Forest, scarce; (rare in Essex); hedges near Crawley; pasture behind Swan Inn, Bell bar; Berry Wood, Aldenham; at Northaw, by the roadside; between Leggatts and the Ridgeway to Tolmers; Brickendon Wood.

SETARIA VIRIDIS.—Cornfields, and about mills, &c.; rare, 7–8. Bexley Heath* (E. de C.); railroad station, Springfield* (E. de C.); about Weybridge; gasworks, Hertford; Watford; cornfields below Buckland Hill; about Puttenham; Brook Farm, near Albury; near Guildford.

SHERARDIA ARVENSIS.—Cornfields and pastures in a light gravelly soil; frequent, 4–10. About Croydon;* fields between Sutton, Carshalton, and Banstead Downs;* between Leatherhead and Epsom Downs;* cornfields above and below the downs, in many places between Guildford and Wrotham;* also about Dartford;* Harefield;* (more frequent apparently on chalk grit, or on gravel over chalk than in alluvial soil, or on gravel over clay.)

SILAUS PRATENSIS.—Pastures and meadows; frequent, 6–9. (Plentiful about Wood Green;* Hendon;* and banks of the Paddington Canal.)

SILENE INFLATA.—Pastures and roadsides; frequent, 6–8. More general in the chalk districts; about Croydon;* roadside between Carshalton, Sutton and the Banstead Downs;* Mickleham, Dorking chalk-pit;* of frequent occurrence on banks and in fields along the base of the downs;* about Dartford; and on the road thence to Darent Wood.*

SILENE ANGLICA.—Sandy and gravelly fields, 6–11, not common. Telegraph Hill, near Ditton;* about Dorking; Weybridge &c.; Duppa's Hill, Croydon; Coombe Wood, cornfields near; Albury; fields W. of Woking, and between Woking and Whitemoor Common; about Frensham; near Wellington College; field right of lane leading from Reigate Heath to Wonham; Dorking; fields about Witley Common; between Hertford and Welwyn.

Silene italica.—Rare, local, 6–7. Darent Wood, and roadside near.

SILENE NOCTIFLORA.—Rare; cornfields in a sandy soil, 7–8. About Broomfield; Hoddesdon.

SILENE MARITIMA.—Local. Southend.

Silybum Marianum.—Banks and waste places; rare, 7. Roadside, in a hedge near Greenstreet Green, coming from Dartford;* by the Thames, near Erith;* Purfleet; Brentwood; Fyfield; road from Hertford to Ware; field between Ball's Wood and Gallows Plain.

SINAPIS ARVENSIS.—Cornfields, common, 6–7. In almost every cornfield.

SINAPIS ALBA.—Waste places, not unfrequent, 6–7. About Epping, and in cornfields on the Hog's Back.*

F 2

SINAPIS NIGRA.—By hedges and in waste places, 6–9, not very frequent. Plentiful on the clay bank, between Leigh and Southend, by the Thames.*

SISON AMOMUM.—By hedges and roadsides; frequent, 8–9. Abundant about Brondesbury;* Harlesden and Willesden;* between Dorking and Ranmore Common; of constant occurrence, and in every description of locality.*

SISYMBRIUM OFFICINALE.—Waste places, common, 6–7. Everywhere in the outskirts.*

SISYMBRIUM IRIO.—Waste places about London; rare, 7–8. Near the gasworks, Hertford.

SISYMBRIUM SOPHIA.—Waste places, among rubbish; rare, 6–8. Box Hill (?); about Erith; Northfleet; Harefield; near Weybridge; Grays Thurrock; Purfleet; Tilbury; between Stanstead and Ware, and road to Ware Park; gravel pit by road to Wotton, Herts.

SISYMBRIUM ALLIARIA.—Hedgebanks and waste places; common, 5–6. Roadside hedges everywhere in the environs. (Tottenham;* Neasdon;* &c.)

SIUM LATIFOLIUM.—Riversides, ditches and watery places; rare, 7–8. Thames, near the reservoirs, Barnes, one large root; near Weybridge, a plant or two;* ditch between Ditton, and Ditton Green; by the Roding; in the river, by St. Mary Cray; marshes below Woolwich; river at Harefield (?); about Northfleet; Merstham pools, between Rotherhithe and Deptford; Sonning, Berks; ponds at Wargrave, and foot of Winter Hill, Berks; Thames, Middlesex side between Twickenham and Richmond.

SIUM ANGUSTIFOLIUM.—Ditches and slow streams, not unfrequent, 7–8. In the Colne, between Rickmansworth and Uxbridge, and in ditches by the towing-path;* in the Cran at Babe Bridge;* ditches in the Lea meadows;* canal at Greenford.

SMILACINA BIFOLIA.—Rare, 5–6. Caen Wood, Highgate.

SMYRNIUM OLUSATRUM.—Waste ground, generally near the sea; rare, 4–6. Dorking chalk-pit; road from Mickleham to Dorking; near Dartford; Stone chalk-pit;* Greenhithe chalk-pit, scarce;* Northfleet chalk-pits, plentiful;* about Gravesend; between Uxbridge and W. Drayton (?); about Rochester; Purfleet; Tilbury.

SOLANUM NIGRUM.—Waste places, also on cultivated land, common, 6–11. Everywhere in the environs. (Hampstead Heath, near;* waste ground about Brondesbury; about West-end, Hampstead;* fields, Putney, Fulham, Tottenham &c.*)

SOLANUM DULCAMARA.—Hedges and thickets; common in cultivated districts, 6–8. (Hedges about Hendon;* Kingsbury;* Tottenham;* Woodford;* &c.)

SOLIDAGO VIRGAUREA.—Woods and thickets, common, but rare on the chalk, 7–9. Hampstead Heath, among the Fern;* on all the Surrey heaths;* roadside from Weybridge to Chobham Common;* Bucks. heaths;* Epping Forest;* Leith Hill;* Holmwood Common;* Chislehurst Common;* woods at Warley; Keston Common; Ball's and Bayford woods; Broxbourne and Wormley woods; Harefield and Pinner woods; Oxhey Wood.

SONCHUS ARVENSIS.—Cornfields and by river-banks; frequent, 8–9. In almost every cornfield; (cornfields at Willesden;* by the rivulet, Wimbledon Common.*)

SONCHUS OLERACEUS.—Both on waste and on cultivated ground, common, 6–8. Everywhere in the suburbs.

SONCHUS ASPER.—In similar situations, and as common, 6–8. Everywhere, and with the preceding.

SONCHUS PALUSTRIS.—Ditches by the Thames, formerly, very rare; now probably extinct. (May be sought for in the neighbourhood of Barking; Rainham; Plaistow; Plumstead.)

SPARGANIUM RAMOSUM.—Ditches and stagnant water, common, 7. (Abundant in the Roding,* ditches, &c., by the Lea.*)

SPARGANIUM SIMPLEX.—Ditches and stagnant water, in a gravelly soil, frequent, 7. In the rivulet in the hollow below Wimbledon Common;* Paddington Canal, beyond Willesden;* Thames about Teddington;* ditches near Waltham Abbey; Colney Heath; Totteridge Green.

SPARGANIUM AFFINE and MINIMUM.—Rare. Two localities in Herts; pond near Digwell's Lodge Farm; and pond S.E. corner of wood at Darman's Green; uppermost of three ponds between St. George's Hill and the Wey? *olim.*

SPARGANIUM NATANS.—*See* S. AFFINE and MINIMUM.

SPARTINA STRICTA.—Sheppey (mouth of the Medway?), local, abundant, 8.

SPECULARIA HYBRIDA.—Dry and chalky cornfields, not common, 6–9. Cornfields near Dartford, on the road to Greenstreet Green;* cornfields on the downs about Reigate;* near Harefield; Brockham; about Croham Hurst; about Verulam, St. Albans; Purfleet; Tilbury; Southend.

SPERGULA ARVENSIS.—Cornfields, frequent, 6–8. (Plentiful about Weybridge;* and at Ham near Richmond.)

SPERGULARIA RUBRA.—Gravelly and sandy soils, frequent, 6–9. (Hampstead Heath;* Plumstead Common.*)

SPERGULARIA NEGLECTA.—Banks of the Thames below Woolwich, local, 6–8. Plentiful on both sides of the river.*

SPERGULARIA MARINA.—*See* S. NEGLECTA.

SPIRÆA ULMARIA.—Ditches and riversides, common, 6–8. (Plentiful by the Thames,* Lea, &c.*)

SPIRÆA FILIPENDULA.—Chalk downs, frequent, 6–7. Purley Downs;* roadside between Croydon and Sanderstead;* Coulsdon;* Mickleham;* Leatherhead;* Epsom and Banstead Downs;* Box Hill;* Reigate Hill;* and in other places along the chalk range;* also by the Thames, incidentally, from seed brought from a distance by floods (?); Morant's Court Hill; chalk downs about Wrotham; about Aldbury Nowers Wood and heath S. of Tring (*olim*, no heath there now).

SPIRANTHES AUTUMNALIS.—Dry hilly pastures, in a chalky or gravelly soil, not common, 8–9. Banstead Downs; Purley Downs; downs about Coulsdon; Reigate Hill;* Mickleham Downs; Dartford Heath; Box Hill; Betchworth Hill; sandpits about Woolwich (?) *olim*; Epping Forest, near 8th milestone (?) *olim*; open pastures about Hanwell; Warley Common (?); Tunbridge Wells Common, and elsewhere in the neighbourhood; Bedwell Park, Essendon; near the old mill, Hoddesdon;

Box Moor; No-man's-land; between Aston and Sheephall; Hatfield Park; field near Brickendon; pasture with *Erythræa pulchella* (which see); Totteridge Green; field near Mill Hill; Pinner Hill.

STACHYS BETONICA.—Woods and thickets, frequent, 6–8. (Hampstead Heath;* Barnes Common.*)

STACHYS PALUSTRIS.—River banks, and moist places, frequent, 7–8. (Plenty by the Lea river;* and by the canals.*)

STACHYS SYLVATICA.—Woods and hedges, frequent, 7–8. (Lanes about Willesden;* Hendon, &c.*)

STACHYS ARVENSIS.—Dry cornfields, not common in the metropolitan districts, 4–10. Fields near Croham Hurst;* cornfields at the foot of the Reigate, Buckland,* and Betchworth hills; and W. of Reigate Heath; and below hills E. of Merstham; near Rickmansworth, in a field between Long Valley Wood and Watford road; cornfields, Harefield; cornfields about Coombe Wood; Fyfield; Southend; cliffs towards Leigh; in the Weald; cornfield about Oak of Honour Wood; fields on Clement's Farm, Brickendon; field on Barber's Lodge Farm, near Hatfield Woodside; cultivated ground within Milward's Park, Hatfield; gardens, Pinner; fields about Hatfield woodside; Broxbourne.

STATICE LIMONIUM.—Muddy shores, and salt-marshes, by the Thames, frequent, 7–9. Between Woolwich and Erith;* Dartford;* Greenhithe;* Northfleet;* Gravesend;* by the Medway at Rochester; Tilbury;* and onwards to Southend.*

STATICE BAHUSIENSIS.—Local, rare. Purfleet.

STELLARIA AQUATICA (syn. *Malachium; Cerastium*).—Watersides, and damp localities, not common, but plentiful where it does occur. Watery places about the Colne and canal at Harefield;* Uxbridge;* by the Cran at Hanworth Bridge;* by the Roding; Barking; Rainham; Ongar; Lea Valley, frequent; Colney.

STELLARIA MEDIA.—Roadsides, waste places, and in gardens, common, 1–12. Everywhere.*

STELLARIA HOLOSTEA.—Woods and hedges, frequent, 4–6. In almost every hedge in the lanes about London; (Tottenham;* Hendon, &c.*)

STELLARIA GRAMINEA.—In similar situations, and on heaths (in dry places), 5–8. (Lanes about Chingford;* Tottenham;* &c.)

STELLARIA GLAUCA.—Marshy places, margins of ponds, &c., not common, 5–7. In the ravine, Wimbledon Common;* Reigate Heath (bogs);* marshy meadows by the Wey near Guildford; between West Ham and the Thames; near Hoddesdon, in a marsh N. of it; Hertford Heath, opposite entrance to Haileybury; and marshy ground S. of old mill; side of ditch in a meadow near Woodbridge railway station.

STELLARIA ULIGINOSA.—Ditches, rivulets, and bogs, frequent, 5–6. (Hampstead Heath; in the bog.*)

SYMPHYTUM OFFICINALE.—Ditches, river banks, and watery places, common, 5–6. (Ditch by the Thames on the Surrey side of Hammersmith Bridge;* also by the river; by the Lea, and in bordering ditches, abundantly.*)

SYMPHYTUM TUBEROSUM.—Shady woods, and river banks, rare, 6–7. Wimbledon Park; near Barnet.

SUÆDA MARITIMA.—Thames shores, local. Grays; Purfleet;* Tilbury;* Southend.*

TAMUS COMMUNIS.—Hedges and thickets, frequent, 6–7. Epping Forest;* lanes about Kingsbury;* and everywhere in woodland localities, occasionally.*

TANACETUM VULGARE.—Waste places, and river sides, frequent, 7–9. Rare in Herts. Banks of the Lea Canal, plentiful;* ditch Edmonton, near Angel Road station, Hoddesdon, &c.*

TARAXACUM OFFICINALE.—Pastures, waysides, and waste places, common, 4–7. Everywhere about London;* the var. *palustre*, by the Thames between Hammersmith and Kew; and on Mitcham Common.

TAXUS BACCATA.—Upland woods, mostly on the chalk and in churchyards (planted); frequent. Farthing Downs (borders);* Box Hill;* Hog's Back;* hills E. of Wrotham;* and other localities on the chalk range;* such as hills W. of Dorking, &c.;* Epping Forest; Great Warley.

TEESDALIA NUDICAULIS.—Sandy and gravelly banks and commons, not very frequent, 4–6. Barnes Common, left by the road across it from Hammersmith;* Esher Common (on Winter Downs);* Putney Heath;* about Hampton Court, in the Park; Epping Forest, in a gravel pit; Ilford, in a gravel pit near; between Hersham and Weybridge; Albury and Shiere heaths; Reigate Heath;* Mousehill Heath, Godalming; Witley Common.

TEUCRIUM BOTRYS. Local, 7.—Box Hill, in a valley right of the road from Mickleham to Headley (plentiful in 1875);* about Bagdon Hill, Bookham; Sanderstead Downs (?).

TEUCRIUM SCORODONIA.—Woods, banks, and heaths, common, 7–8. (Hampstead Heath; plentiful.*)

THALICTRUM FLAVUM.—Banks of rivers and ditches, in wet meadows, not common, 6–7. In the meadows between Chingford and Ponder's End, but not frequent;* by the Thames between Walton and Weybridge;* by the Colne at Uxbridge;* by the Brent at Greenford; by the Roding near Chigwell;* by the Cray; by the Colne, between Harefield and Rickmansworth; by the Lea in several places; by the Wey.

THESIUM HUMIFUSUM.—Elevated chalky banks, and pastures, rare, 5–7. Banstead Downs (?); Mickleham Downs; about Coulsdon * (chalky bank between Coulsdon Heath and Kenley Common); Box Hill; Betchworth Hills; between Dorking and Ranmore Common; downs about Guildford; Hog's Back; Purley and Sanderstead downs; Norbury Park, plentiful.

THLASPI ARVENSE.—Fields and roadsides, not unfrequent, 5–7. Cornfields near Willesden;* and about Ongar; foot of Reigate and Buckland hills.*

THYMUS SERPYLLUM.—Hilly and dry pastures, common, 6–8. Everywhere on the downs;* roadside between Carshalton and Banstead.*

THYMUS CHAMÆDRYS.—A variety of the above, perhaps in damper and cooler situations than affected by the typical form. ("Printed words being very satisfactory distinctions, if Nature would only act up to rules, instead of going perversely contrary to printed text." 'Cybele,' Compendium.) Warley Common; Southend.

TILIA GRANDIFOLIA.—Woods and hedges, rare, 6–7. Banks of the Mole near Dorking;* near Box Hill.

TILIA PARVIFOLIA.—Woods in Essex, not common, 7–8. Epping Forest; Ashtead Park, Epsom; Purfleet.

T. intermedia (syn. *europæa*).—Plantations, 7. Frequent about London, in parks, &c.

TORDYLIUM MAXIMUM.—Rare. Waste ground about London and Eton, formerly; found by the author on ditch banks, Tilbury, in 1875, in considerable abundance (yes: H. C. Watson).

TORILIS INFESTA.—Hedges and waysides, not uncommon, 7–9. About Moulsey; between Box Hill and Dorking; Smitham Bottom;* Sunbury; frequent on the chalk range.

TORILIS NODOSA.—Waste places, and by roadsides, especially in the chalk districts, frequent, 5–7. In profusion on banks by the Thames below Woolwich towards Erith.*

TORILIS ANTHRISCUS.—Hedges and waste places, common, 7–9. Roadside hedges and lanes about London, everywhere.*

TRAGOPOGON PRATENSIS.—Meadows, pastures, and waysides, frequent, 6–7. Plentiful by the towing-path, Lea Canal.*

Tragopogon porrifolius.—Moist meadows, very rare, 5–6. Incidentally on railway banks at W. Drayton; at Croydon;* formerly abundant in the marshes below Woolwich; but extinct since these have been drained; still, may be sought for and possibly found about Greenhithe; Erith; Purfleet; &c., Greenford (*olim*).

TRIFOLIUM SUBTERRANEUM.—Dry gravelly pastures, not very common, 5–6. Moulsey Hurst,* in some abundance; Wimbledon Common (scarce);* Ham Common;* Ditton Green; Uxbridge Moor; Dartford (? heath); Shirley Common; Streatham Common; Hampton Court (in the park?); perhaps in Bushy Park; Warley Common; Box Moor; Redhill; Wray Common; (rare in Herts;) Sand-pit Lane, St. Albans; Bernard Heath; common by Walton Bridge.

TRIFOLIUM PRATENSE.—Meadows and pastures, common, 5–9. Generally in meadows, but not in all of them, about London.*

TRIFOLIUM MEDIUM.—Pastures, frequent, but much less so than the preceding species, 6–9. About West-end railway station, Hampstead, sparingly;* about Reigate in pastures.

TRIFOLIUM OCHROLEUCUM.—Pastures and waysides on chalk or gravel, not common, 6–8. Near Abbott's Roding; in close, by fish-ponds, Heron Gate; Great Warley; High Rocks, Tunbridge Wells; between Hertford and Bayford; in a meadow by a brook on the footpath E. of Bayfordbury; also half a mile nearer Hertford; about Sawbridgeworth; frequent about Thorley and Shenley, Herts.

TRIFOLIUM MARITIMUM.—Salt-marshes, local, 6–7. Erith; Greenhithe; Tilbury;* Northfleet and Gravesend; Rochester.

Trifolium incarnatum.—In fields and borders of the same where the plant has been raised as a crop; fields near Reigate Hill.*

TRIFOLIUM ARVENSE.—Cornfields and dry pastures; fallow-fields, &c., frequent, 7–9. Box Hill; Banstead Downs; fallow field near Croham Hurst;* cornfields generally,* but not abundant.

TRIFOLIUM STRIATUM.—Fields and dry pastures, frequent, 6–7. Dartford Heath;* Smitham Bottom;* about Gravesend; Blackheath, road crossing it towards Morden College; Park Hill; Wray Common; many places about Hertford; Hertingfordbury; Bayford, Essendon and Hatfield; &c., Oliver's Mount; Uxbridge Moor; roadside between Staines and Hampton; Mitcham Common; gravel pits, Moulsey Hurst; Shiere Heath; Redhill and Reigate Heath.

TRIFOLIUM SCABRUM.—Chalky and dry sandy fields, not common, 5–7. Generally by or near the sea; Gravesend; Dartford; 2 miles beyond Tilbury; Southend and Shoebury; Mitcham Common.

TRIFOLIUM GLOMERATUM.—Gravelly heaths and pastures, not common, 6. Blackheath and Greenwich Park (?) *olim*; Greenhithe; grassy bank, near Moulsey Hurst, by road from Hampton Court Bridge to W. Moulsey.

TRIFOLIUM REPENS.—Meadows and pastures, common,* 5–9. Everywhere, Wandsworth Common; Edgware Road;* &c.

TRIFOLIUM FRAGIFERUM.—Meadows and pastures, not unfrequent, 7–8. On the Edgware Road, by Child's Hill;* in Epping Forest; below Purfleet; Tilbury; frequent in Herts; Colney Heath; North Marsh, Hoddesdon.

TRIFOLIUM PROCUMBENS.—Dry pastures and borders of fields; frequent, 6–8. Everywhere about London.*

TRIFOLIUM MINUS.—In similar situations, often by roadsides, frequent, 6–9. Roadside between Willesden and Kingsbury, &c.

TRIFOLIUM FILIFORME.—Dry pastures chiefly near the sea, rather uncommon, 6–7. Wimbledon Common;* Wandsworth Common; near Erith; Keston Common; Epping Forest; Hampstead Heath.

TRIGLOCHIN PALUSTRE.—Wet meadows, and by the sides of rivers, and ditches in marshy situations, frequent, 6–8. Between Greenwich and Woolwich; Dartford Marshes; marsh-ditches below Erith in abundance;* boggy meadow in Wormley Wood;* bogs, Gomshall Common; marsh near Redhill railway station; Colney Heath; about Rickmansworth; Harefield, &c.

TRIGLOCHIN MARITIMUM.—Salt-marshes, local, frequent, 5–9. Plentiful by the Thames below Woolwich on both sides of the river;* Purfleet;* Tilbury;* Northfleet;* &c.

TRIGONELLA ORNITHOPODIOIDES.—Dry, sandy pastures, rare. Hanwell Heath (?); Northaw by the road to Goff's Oak, Herts; also near Coffleys, Herts; Wandsworth Common (?); Wimbledon Common (?).

TRIODIA DECUMBENS.—Heaths and moors, frequent, 7. Hampstead Heath;* Coulsdon; on the Surrey heaths in moory parts; Whitemoor;* Bagshot;* Chobham;* Leith Hill;* Pirbright Heath;* and in Epping Forest;* Warley Common.*

TRITICUM CANINUM.—Woods and banks, frequent. (Hedgebanks about Willesden; &c.*)

TRITICUM REPENS.—Fields and waste places, 6–8. Everywhere in the outskirts.*

TRITICUM JUNCEUM.—By the Thames below Gravesend, local, abundant, 7–8. Tilbury;* Cuxton;* Southend;* Canvey Island.*

TULIPA SYLVESTRIS.—Pastures, very rare, 4. Wimbledon Park;* in a triangular field at junction of Kingston and Merton Road; in an orchard near railway bridge, Buckland; orchard at Egypt, Holmwood Common.

TURRITIS GLABRA.—*See* ARABIS PERFOLIATA.

TUSSILAGO FARFARA.—Moist and clay soils, common, 3–4. Banks of the Lea Canal;* between Finchley Road and West-end; Hampstead railway station.*

TYPHA LATIFOLIA.—Borders of ponds and rivers, frequent, 7–8. In the Lea;* Roding;* ditches by the Thames in the marshes.*

TYPHA ANGUSTIFOLIA.—Ditches and pools, not uncommon, 7. Ditches by the G. E. Railway between Clapton and Broxbourne, in several places;* ditches in the flats below Plaistow onwards;* Canvey Island;* Tring reservoir; Ruislip reservoir; canal between Hanwell and Brentford; ponds, Wandsworth Common; Weybridge; pond near Sunbury Lock; Gatton Pond; forked ponds, Witley.*

ULEX EUROPÆUS.—Heaths and commons, frequent, 2–7. (Hampstead Heath;* Barnes Common.*)

ULEX GALLII.—Dry heaths, frequent, 7–11. Ditton Common;* Keston Common;* on all the heaths in S.W. Surrey; abundant,* and on the dry Buckinghamshire heaths to the exclusion almost of the ordinary kind,* Stoke Common;* Farnham Common; by Burnham;* Gerard's Cross Common;* No-man's-land Common; Colney Heath.

ULEX NANUS.—Old name, now applied to the stunted prostrate form, in a *specific* sense. No-man's-land (T. B. Blow).

ULMUS SUBEROSA.—Woods and hedges, common, 3–5. By the Edgware Road; &c., in Hyde Park, plentiful.*

ULMUS MONTANA.—Woods and hedges, frequent, 3–4. In the parks.*

URTICA DIOICA.—Waste places and hedge-banks, common, 6–9. Everywhere in the suburbs.

URTICA URENS.—In similar situations, and as common, 6–9. With the preceding, but generally in less shady localities.

Urtica pilulifera.—Under walls, and among rubbish, near habitations, 6–8. Rare, casual.

UTRICULARIA VULGARIS.—Ditches and ponds, rather unfrequent, 6–7. Pools and holes near the Brookwood station on Pirbright Heath;* in Heron Gate pond, Thorndon (?) *olim;* pond near Egham; ponds, foot of Cookham Down; ponds, Felbridge; pond on Epping Forest; between W. Ham and the Thames; near Sonning; ditches, Stanstead; Hoddesdon Marsh, first ditch above the new mill; pond in a thicket by Pembridge Lane, near Broxbourne Lodge.

UTRICULARIA INTERMEDIA.—Rare. Ponds, Burnham Beeches.

UTRICULARIA NEGLECTA.—Confused with the preceding. Gravel-pit, Hainault Forest.

UTRICULARIA MINOR.—In similar situations, very rare, 6–9. Peaty bogs on Farnham Common; river at Uxbridge, (?); in the Basingstoke Canal, opposite Woking station; Pirbright Common, between the canal and railway; turf-pits, Thursley Common.

VACCINIUM MYRTILLUS.—Woods and heaths in hilly countries, 4–6. Abundant where it occurs; Leith Hill;* Hurtwood Common;* hilly parts of Pirbright Heath;* Chobham ridges;* also on Croham Hurst;* and on Hampstead Heath; near the 'Spaniards;'* Epping Forest; in one place; Oxhey woods.

VACCINIUM OXYCOCCOS.—Rare, local, 6. In a bog near Brough Farm by Witley Common (?); not seen in 1876; the bog in part drained, and in another part the turf cut up for peat; spongy places, Black Lake, near Farnham; a patch in Epping Forest (?) *olim*; very doubtful.

VALERIANA DIOICA.—Marshy meadows, frequent, 5, 6. Ruislip, moory meadows;* in a ravine on Wimbledon Common;* Warley Common; Hackney Marshes (?); by the Roding; Epping Forest; near Ongar; Harefield.

VALERIANA OFFICINALIS.—Ditches sides of rivers, moist woods, frequent, 6–8. By the Colne;* in ditches by the Thames; Surrey side;* near Godalming;* Chislehurst woods;* Wormley Wood; Epping Forest.* Var. *sambucifolia;* bog foot of Frith Hill, Godalming.

VALERIANELLA OLITORIA.—Banks and cornfields in a light soil, frequent, 4–6. Fields between Sutton and Banstead Downs;* about Woodford;* Willesden;* Ditton;* and in most cornfields and bordering hedges.

VALERIANELLA DENTATA.—Cornfields and hedgebanks, not very common, 6–8. Cornfields about Telegraph Hill, Ditton;* cornfields on the Sutton Downs, and on Reigate Hill; Box Hill; &c., Croham Hurst, Croydon; Broomfield. Var. *mixta* in a field behind lock keeper's house, Teddington (Henfrey); fields near Norton Heath; Ruislip.

VALERIANELLA AURICULA.—Rare. Claygate; and near Chessington Church; about Hook; cornfield, foot of Box Hill (S. side); between Guildford and St. Martha's Hill.

Valerianella carinata.—Near Ditton (H. C. Watson); near Ongar (Lindley); a casual; at Marden, Ash.

VERBASCUM THAPSUS.—Banks and waste ground in a light, sandy, gravelly, or chalky soil, frequent, 6–8. Lanes about Colney Heath, plentiful;* about Gravesend; Erith; Dartford;* Greenhithe;* Box Hill;* Epping Forest; Purfleet; Great Warley.

VERBASCUM PULVERULENTUM.—Roadsides on a chalky or gravelly soil, rare, 7. Between Guildford and Shalford (?) *olim* (none there now); near Cuxton *olim*.

VERBASCUM LYCHNITIS.—Roadsides, pastures and fields in a chalky soil, 7–8. Rare; between Dartford Heath and Darent Wood; also between Dartford and Greenstreet Green, pretty plentifully;* between Guildford and Shalford, and on Shalford Common (?) *olim* (none there now); Chislehurst Common (?); between Bromley and Chislehurst;* Croham Hurst; Smitham Bottom; Cobham; N. side of Morant's Court Hill; chalk-pit, Riddlesdown; abundant* (E. de C.); railway station, Bickley* (near it); between Erith and Greenhithe; Wargrave, Berks.

VERBASCUM NIGRUM.—Banks and waysides in a gravelly or chalky soil, frequent, 6–10. About Croydon,* towards Smitham Bottom; Box Hill* (plentiful); between Dorking and Brockham; between Guildford and Shalford;* near Godalming;* about the ruins of Verulam, St.

Albans; roadsides about Farnham; about Harefield; Purfleet; near Shoebury; Hoddesdon; Sawbridgeworth; Watford; Rickmansworth; field near N.E. boundary of Hatfield Park.

Verbascum Blattaria.—Banks in a gravelly soil, rare (in Kent), 6–10. Mitcham Common, near Hackbridge (?); between Crayford and Dartford (?) *olim*; chalk-pits near Rochester; about Cobham (?); Binfield, Berks.

VERBASCUM VIRGATUM.—Fields and roadsides, rare, 8. Between Guildford and Shalford? and on Shalford Common (?) *olim* (none there now); Great Berkhampstead Common (?).

VERBASCUM THAPSO-NIGRUM, (THAPSIFORME).—Rare, 7–8. About Dartford; Charlton and Greenhithe (?) *olim*; about Box Hill and Mickleham; near Cuxton, *olim*.

VERBENA OFFICINALIS.—Roadsides and waste ground, frequent, 7–9. About Guildford;* Croydon;* Shorne;* Cobham;* bordering banks &c., on the Downs in many places;* by the Thames; between Fulmer and Iver Heath;* Leatherhead and Mickleham;* Esher;* roads between Dartford, Dartford Heath, and Greenstreet Green;* Kingston Road; Wimbledon Common, a few plants;* outskirts and bordering lanes, Epping Forest;* &c.

VERONICA HEDERIFOLIA.—Fields and hedgebanks, common, 3–8. Hedgebanks in lanes, Child's Hill;* Hendon;* &c.

VERONICA POLITA.—Fields and waste places, common, 4–9. Everywhere in the environs.*

VERONICA AGRESTIS.—In similar situations, common, 4–9. Everywhere in the environs.*

VERONICA BUXBAUMII.—Cornfields, frequent, 4–9. In almost every cornfield, especially in the chalk districts;* near Harlesden Green;* Acton;* Sutton Downs;* Essex cornfields.*

VERONICA ARVENSIS.—Fields and walls, common, 4–7. Fields, especially such as are fallow, and in unweeded market-gardens in the environs, everywhere.*

VERONICA SERPYLLIFOLIA.—Pastures and roadsides, common, 5–6. Hampstead Heath;* Hendon, &c.*

VERONICA OFFICINALIS.—Woods and pastures, frequent, 5–7. Pinner and Oxhey woods;* abundant.

VERONICA CHAMÆDRYS.—Woods, pastures and hedgebanks, frequent. Everywhere in hedgebanks, well away from the bricks and mortar, Hendon;* &c.

VERONICA MONTANA.—Moist woods, not common, 4–7. Wood near Breakspeares, Harefield; near Croydon, (? Croham Hurst); woods, Wimbledon; about Coulsdon; Fridley Copse, Mickleham; woods about Merstham; Gatton; Chipstead; Epping Forest; woods about Godalming; woods S. of the Lea; Old Park Wood, Harefield; and hedgebank near, on the Ruislip road; Albury; Shiere; foot of Leith and of Boar hills.

VERONICA SCUTELLATA.—Wet places and sides of ditches, frequent, 7–8. Putney Heath;* Keston Common;* Epping Forest;* ponds, &c., Warley Common; Esher Common.

VERONICA ANAGALLIS.—Ditches and watery places, frequent, 7–8. Plenty by the Thames; between Hammersmith and Mortlake.

VERONICA BECCABUNGA.—Ditches and watercourses, common, 5–9. In the ditch on Hampstead Heath;* plentiful, (E. or S. heath).

VIBURNUM LANTANA.—Woods and hedges on a chalky soil, abundant, 5–6. Everywhere on the chalk range from the Hog's Back to Cuxton;* between Crayford and Dartford; between Dartford and Darent Wood;* and in the wood;* about Greenhithe and Northfleet;* Purfleet;* Harefield;* Tring;* lanes in Essex, between Hatfield Broad Oak and Fyfield, occasionally (gravel over chalk at some depth);* about Hertford; St. Albans; &c.

VIBURNUM OPULUS.—Woods and coppices in damp situations, as well chalky as otherwise, 6–7. Box Hill; Darent Wood;* Stroud Copse, Godalming; Epping Forest;* by the Brent about Totteridge, plentiful.

VICIA CRACCA.—Bushy places, frequent, 6–8. Thames bank above Teddington;* Hampstead Heath, E.;* Box Hill, &c.*

VICIA SATIVA.—Cultivated borders of fields where a crop of the plant has been previously raised, 5–6. Between Wimbledon stat. and Morton, near Mitcham.*

VICIA ANGUSTIFOLIA.—Dry pastures in a sandy or gravelly soil, rare, 5–6. Darent Wood; about Eltham, (?) *olim.*

VICIA HIRSUTA.—Hedges and cornfields, frequent, 6–8. Epping Forest; between Harlesden, Willesden, and Neasdon;* Warley.

VICIA TETRASPERMA.—Cornfields, hedges, and bushy places, frequent, 6–8. Bushy places and hedges by the Paddington Canal;* bordering banks in the marshes below Woolwich, plentiful.*

VICIA SEPIUM.—Woods and shady places, frequent, 6–8. Epping Forest;* Darent Wood;* Chislehurst;* about Charlton; roadside between Fulmer and Iver Heath;* about Hendon,* &c.

VICIA LATHYROIDES.—Roadsides and dry pastures, not unfrequent, 4–6. Epping Forest;* Esher; banks about Greenhithe.

VICIA BITHYNICA.—Bushy places in a gravelly soil, rare, 7–8. Darent Wood; near Hadleigh Castle.

VICIA SYLVATICA.—Woods, rare. Woods above Aston Hill, Tring (?) *olim.*

VILLARSIA NYMPHÆOIDES.—*See* LIMNANTHEMUM.

VINCA MINOR.—Hedges and banks in woods, not frequent, 4–6. Wimbledon Common (?); Epping Forest, near Woodford; and in Larkswood, Chingford; lane between Swanscomb and Darent Wood; lane near Harrow Weald Common;* Westhumble Lane, near Burford Bridge;* Hatfield Broad Oak; Fyfield; thicket, South Weald, Little Warley; near Croydon; Cobham (Sur.); one mile from Mary Cray, on the road to Chelsfield; near Totteridge; Theydon Bois; Hertingfordbury Park and Mole Wood; Boxwood, Hertford Heath; Wix's Wood, Tring; lane near the Union, St. Albans; lanes at King's Langley; Bentley Priory woods; by the Engine Pond, Gatton; about Albury; Mickleham; Clandon.

Vinca major.—Woods and thickets, rare. Hedge on Esher Common; Woodford; Coulsdon.

VIOLA PALUSTRIS.—Bogs and marshy places, not common, 4–7. Hampstead Heath;* marsh at the foot of the hills E. of Merstham;* Sunninghill Bog, near Ascot;* Farnham;* common near Burnham Beeches;* Warley Common; meadow near Shirley Common; about Dorking; Broadmoor, lower extremity; ravine, Leith Hill, towards Wotton.

VIOLA ODORATA.—Hedges and woods, frequent, 3–4. Banks about Kingsbury,* Mortlake,* Burford Bridge,* Croydon,* &c.

VIOLA HIRTA.—Hedges, copses, and banks in the chalk districts, frequent, 4–5. Box Hill, Betchworth, and in many places along the chalk range;* Harefield;* Coulsdon; Tring;* Dartford, road to Greenstreet Green;* Banstead Downs* (var. *calcarea*?); Epping (?); Southend (?); Hog's Back; Harefield.*

VIOLA SYLVATICA (*olim* CANINA).—Woods, banks, and dry pastures, frequent, 4–8. Plentiful in Epping Forest.*

VIOLA CANINA.—With var. *flavicornis* in similar situations; probably not very unfrequent, 4–8. A mere variety of the preceding; between Hampton and Staines.

VIOLA LACTEA.—In similar situations (?), rare, 4-8; a variety only of the last. Roadside, from Bagshot to Ascot, in an old brickfield. N.B. The student should gather these Dog Violets from various localities, and compare them one with another, and with the descriptions in the books (*vide* observations in 'Cybele Brit.' Compend. p. 441, on this species). Records of localities are wanting for them, except as an aggregate.

VIOLA TRICOLOR.—Cultivated fields and banks, frequent, 5–10. In almost every cornfield, and in fallow fields where corn has grown.*

(VIOLA LUTEA.—No record ?)

VISCUM ALBUM.—On trees, but not very frequent in the metropolitan districts. In the Park at Betchworth; in Norbury Park; Epping Forest, near Chingford and Loughton; on thorns, about Cobham; in Bayfordbury and Brickendonbury Parks; in Moor Park; on thorns, Bushy Park.

WAHLENBERGIA CAMPANULACEA.—Bogs on the heaths in S.W. Surrey mostly rare, 7–8. By the forked ponds on Witley Common;* bog near High Beech, Epping Forest; Tilgate Forest, a little beyond Starvemouse Moor; Keston Common, margin of lower pool; on moors near Tunbridge Wells.

WOLFFIA ARRHIZA.—Ponds, rare. Pond in a garden, at Byfleet; and at Staines; also in a field S.E. of St. James's Church, Walthamstow.

ZANNICHELLIA PALUSTRIS.—Ditches and stagnant waters, frequent 5–8. About Staines; near Godalming;* pond near Finchley Road station;* ditches by the Lea, and in the Lea Canal, between Rye House and Stanstead; at Hoddesdon, Broxbourne, &c.; Tring reservoir.

ZANNICHELLIA PEDICELLATA.—A variety. Marsh ditches by the Thames below Woolwich.*

ZOSTERA MARINA.—Ditches by the Thames, below Gravesend, local, 7–8 Southend.

II. CRYPTOGAMS.

ASPIDIUM ACULEATUM, scarce.—Moist woods and shady banks, rocky places. Wimbledon Common; about Hendon; Epping Forest; about Muncombe; Old Park Wood, Harefield; Chigwell; Dorking; swampy grounds, N. of Cold Harbour; hedgebanks, Chert Lane, Reigate; near Welham Green and Colney Heath; Warley Common; Pinner Wood and banks near Pinner; between Cobham and Leatherhead; about Earlswood Common, Nutfield and Bletchingley; Brentwood, Little Warley Common; Burnham Beeches.

Var. *lobatum*.—Between Harefield and Ruislip Brook, near Bayford Church; road from Hatfield to St. Albans; from Essendon to Little Berkhampstead; Essendon West-end; roadside near Welham Green and between the green and Colney Heath; Great Berkhampstead; St. Albans; Epping and Hainault forests; Warley Common; about Reigate and swampy wood near Leith Hill; Burnham Beeches.

ASPIDIUM ANGULARE.—Woods and hedgebanks between Colney Heath and N. Mimms; lanes west side of the ruins of Verulam; foot of Leith Hill, N. of Cold Harbour; hedgebanks, Chert Lane, Reigate; Warley Common; banks near Pinner; lane from Pinner to Harrow Weald Common; pits by the road to Hertford and lanes about Hertford; road from Cole Green to Hertingfordbury; Mount Pleasant, Brickendon; lane between Keber Green and Hatfield woodside; between Hatfield railway and the Union Workhouse; lane near Barking; Epping Forest; Burnham Beeches.

ASPLENIUM RUTA-MURARIA.—Old walls, not uncommon. Church at Teddington; wall near Highgate, towards Hampstead;* walls at Bletchingley and Godstone; in Norbury Park, Merrow Downs; Shalford Church; about Albury; Shiere; in Herts, several places; Epping Church; Woodford; Leytonstone.

ASPLENIUM TRICHOMANES.—Rocks and walls, uncommon. Haslemere near Guildford; stony bank, Hammer near Gomshall; old walls between Foots Cray Church and Hurst; ditto near Woodford; lane between Ongar and Kelvedon; about Shiere; hedgebanks about Reigate; between Hatfield woodside and Welham Green; lane from Cashiobury to Rouse farm, and elsewhere in the Colne districts; Warley Common; orchard in Harefield Place; about Breakspeares, Harefield; rock on Tunbridge Wells Common; Walthamstow; Leyton; Rainham Church; Burnham Beeches; walls of Burstow Church; about Little Berkhampstead; Thieves Lane, Hertford; lane between Bayford and Little Berkhampstead; and by a bridle way from this to Westuble Lane; on an old wall at Rickmansworth, between Moor Park and the Colne.

ASPLENIUM ADIANTUM-NIGRUM.—Banks, walls, and fissures of rocks, not common. Haslemere near Guildford; Charlton Wood (?); Darent Wood; Keston Common; old wall at Leytonstone (?); Loughton Church; Brook End (Ess.); Warley Common; Teddington Church; woods and lanes about Shiere and Albury; hedgebank, Harrow Weald Common; Headstone Lane, Pinner; lane leading to Rickmansworth from Harefield; hedgebanks S. of the chalk range in the lower greensand districts, frequent; Thieves Lane, Hertford; between Brickendon Green and Blackfan Wood; between Bayford and Little Berkhampstead; Essendon; between Hatfield woodside and Welham Green; between Hatfield and St. Albans; between St. Albans and Watford; between Rustall Common and Langton Green; Norton Heath; Burnham Beeches; Tilgate Forest.

ASPLENIUM LANCEOLATUM.—Rare; sand rocks, by a path leading from Tunbridge Wells to High Rocks.

ATHYRIUM FILIX-FŒMINA.—Moist places in woods mostly, not frequent, Guildford chalk-pit (?); Darent Wood; Warley Common; Colney Heath (?); Great Berkhampstead Common, near the brick-kiln; copse near Bell Bar (plenty); Berry Wood near Aldenham; boggy copse W. of Reigate Heath;* Oxhey Wood; enclosure near Harrow Weald Common; Winchmore Hill Wood; moist woods S. of the chalk range; Bell Wood, Bayford; Hatfield woodside; Cook's Hill, Little Berkhampstead; Colney Heath; lanes near Hasloc farm, Tring; Fyfield; Snaresbrook; Warley Common; Burnham Beeches; Tilgate Forest.

BLECHNUM.—*See* LOMARIA.

BOTRYCHIUM LUNARIA.—Hilly pastures, downs, &c., meadows, not frequent. Shirley Common; about Chislehurst; Coulsdon; between Dartford and Foots Cray; summit of Leith Hill; Shackleford Heath; Farnham Park; Reigate Heath, N.W. corner; Bury Hill, near Dorking; Puttenham Heath.

CETERACH OFFICINARUM.—Old walls &c., rare. On Cliffe Church, below Gravesend; lane leading from Mickleham to Headley (?) *olim*; in Herts in one place, locality not stated; at Woodford (?) *olim*, on two walls and on a tomb.

CHARA FLEXILIS.—Stagnant water. About Totteridge; Hendon; Epping Forest;* Leatherhead.

CHARA SYNCARPA.—Ponds. (Ruislip reservoir.)

CHARA TRANSLUCENS.—Ponds. (High Beech; Epping Forest; Stanmore Heath.)

CHARA FŒTIDA.—Stagnant water. Pinner; pond in field right of lane leading from Hampstead Heath to Fortune Green;* ditch by the Thames side, opposite Sunbury Lock.*

CHARA HISPIDA.—Ponds. Lake in Hatfield Forest; Ruislip Common; bog in Wormley Wood.*

CHARA TOMENTOSA.—Ponds. (Chislehurst.)

CHARA FRAGILIS.—Ponds. (Pinner Hill.)

CYSTOPTERIS FRAGILIS.—Walls, very rare. At Leytonstone formerly on a wall; in the road leading from Weston Street to Albury Park; and wall opposite Weston House, Albury (? same localities although differently described).

EQUISETUM ARVENSE.—Fields and banks, common. Roadsides, &c. about Neasdon and Kingsbury;* banks of the Lea navigation.*

EQUISETUM MAXIMUM.—Wet damp grounds, sides of ditches, clay banks and swampy bogs, frequent. Wormley Wood;* wet wood near Northaw, S. side of the Ridgway; between Barnet Gate and Totteridge; Rickmansworth; Gatton; below hills E. of Merstham; and in wet places below the hills W. of Reigate to Dorking; wet places near Guildford.

EQUISETUM SYLVATICUM.—Moist woods and hedgebanks, not common. Woods about Chislehurst; Epping Forest; Harefield; Bayford Wood, and Bell Wood, near Hertford; Harrow Weald Common;* between Peslik (near Shiere) and Ewhurst, in a valley; boggy wood on Holmwood Common, and on Boar Hill; swampy woods near Cold Harbour, Leith Hill; boggy thicket on Warley Common; Burnham Beeches.

EQUISETUM PALUSTRE.—Boggy places, common. Hampstead Heath;* Ruislip Moor;* Uxbridge Moor.*

EQUISETUM LIMOSUM.—Lakes, sides of pools and rivers, ditches &c., frequent. Warren pond, near Breakspeares; Harefield; Rickmansworth; Great Berkhampstead; ditches by the Lea canal, especially near Ponder's End.*

EQUISETUM HYEMALE.—Boggy woods, rare. Wanborough Wood, N. side of the Hog's Back, plentiful, two or three acres of it.

HYMENOPHYLLUM TUNBRIDGENSE.—Wet sandy rocks in the Weald of Sussex; rare, local. High Rocks, Tunbridge Wells; in Tilgate Forest; Forest Ridge; on the main ridge, and on Chiddingly Rocks; on a rock called Pook Church in Cow Wood, left of the Brighton road, and half a mile from Handcross; about Ardingly.

LASTRÆA.—*See* NEPHRODIUM.

LOMARIA SPICANT.—Woods and heaths, frequent. Keston Common; Warley Common; Epping Forest; Shirley Common; Leith Hill; Esher Common; Reigate Heath; N. Mimms Wood; Newlands Wood, Rickmansworth; Bacher Heath; Harrow Weald Common; Holmwood Common; Hatfield Heath; Broxbourne woods; Witley Common; Puttenham Common;* Burnham Beeches; Tilgate Forest; and bordering hedges of the Surrey heaths.*

LYCOPODIUM CLAVATUM.—Upland heathy pastures, rare. Leith Hill; Keston Common; Woking Heath, *olim*, (near the station, borders of small water-courses, left of the railway arch?); Shirley Common; Chobham Common; side of pass over the ridge to Frimley, E. side; Witley Common, near Redborough; Tilgate Forest occasionally; Epping Forest, near the King's Oak; between Loughton and Epping (?); heath, Highdown, near Godalming.

LYCOPODIUM INUNDATUM.—Moist, heathy places. Frequent on the Surrey heaths; Wimbledon Common, in the ravine near the camp; Chislehurst Common; Esher Common, near Claremont;* Keston Common;* Burnham Beeches; White Moor Common,* abundantly; Tilgate Forest; Leith Hill Common;* Virginia Water.

LYCOPODIUM SELAGO.—In similar situations, but far less frequent. Shirley

Common; near Woking Common station (?) with *L. clavatum*, *olim*; wet places, foot of Leith Hill; Highdown Hill, near Godalming; E. side of pass, over Chobham ridges to Frimley; Tilgate Forest, left of the road from Tunbridge Wells to Frant; and in abundance on banks of a pond below the bog, between Pease Pottage Gate and Starvemouse Plain, with *L. clavatum* and *L. inundatum*.

NEPHRODIUM FILIX-MAS.—Woods and shady banks, frequent. Epping Forest;* Broxbourne and Wormley woods;* woods S. and S.W. of Hertford; Darent Wood;* Keston Common;* bordering hedges, Surrey heaths;* Burnham Beeches and adjoining woods;* Leith Hill.*

NEPHRODIUM SPINULOSUM.—Boggy heaths and woods, rather rare. Reigate Heath, adjoining wet copses;* Elstead Common, bushy borders of a pond there;* borders of the forked ponds, Witley Common;* lane between Chobham and Bagshot; in two boggy copses, near Hatfield woodside; Winchmore Hill Wood; Horton Wood by Ashtead Common; between Cobham and Oxshott, in a coppice near Little Heath; in a boggy field between Hare Lane and Esher; boggy thickets, Leith Hill districts and Hurtwood; Ball's Wood; Broxbourne woods; bogs about Shiere and Albury; Epping Forest; Warley.

NEPHRODIUM DILATATUM.—Moist woods, moors and shady places, frequent. Lane between Chobham and Bagshot; Reigate Heath; Newlands Wood, Herts; Stanmore Heath; Bartleswell, near Harefield; Whitton Park; Winchmore Hill Wood; Esher; about Shalford and Shiere, plenty; Woking Common; boggy places, Wonham and foot of Leith Hill; and in moist woods between Dorking and Leith Hill; Epping Forest; Warley Common; Oxhey and Pinner woods; Harrow Weald Common.

NEPHRODIUM CRISTATUM.—Boggy heaths, very rare. Burtleswell, near Harefield (?).

NEPHRODIUM THELYPTERIS.—Marshy and boggy places, rare. About Chigwell; Wimbledon Common (?) *olim*, valley near the camp; Epping Forest; near Godalming, in a field, at one spot; Waterdown Forest, Tunbridge Wells; woody bog in a field, near Epping; Keston Common; Windsor Park; Sonning Hill Wells.

NEPHRODIUM OREOPTERIS.—Heaths, and dry pastures in hilly places, not frequent. Epping Forest; Brasted; Leith Hill; Moor Park; Berkhampstead Common, near the brick-kiln; Hurtwood Common; hedgebank bordering a wood under Oxshott Hill; roadside near Long Cross, Chobham Common; near Horsell Common; near Guildford; near Bellwood, Bayford; wood near Northaw, S. side of the Ridgeway, plentiful; Broxbourne Wood, near Well Green; Wormley Wood; near Cow Heath; Shiere Lane, Tring, and in a lane parallel to it; Moor Park; Berkhampstead Common; Warley Common; Witley Common; Holmwood Common, &c.; Harrow Weald Common; Rusthall Common.

OPHIOGLOSSUM VULGARE.—Moist pastures, and in woods, not unfrequent. Darent Wood; about Croydon; Coulsdon; about Cobham, field near the Plough; Downside; Epping Forest; meadows below the hills, E.

of Merstham; and meadows in Buckland and Betchworth parishes; Elstree reservoir; pastures, No-man's-land Farm; wet parts of Albury Park; Waterdown Forest; field at Greenford, plenty; Watford, near the silk mills; Oxhey Wood; Perivale; Harefield, &c.; Loosely Park near Guildford; between High and Chipping Ongar; Great Warley; marshes near Lea Bridge; damp meadows below Reigate Hill; pastures about Bayford, Essendon and Brickendon Green.

OSMUNDA REGALIS.—Boggy places and wet margins of woods, very rare. Devil's Punchbowl, near Godalming; wood W. of Warley Common (?); Burnham Beeches (?) (none there now); Farnham Common, near Cæsar's camp; Leith Hill, above Lonesome; and about Peslik near Shiere; also, other side of the hill, near Ewhurst; Gracious Pond (?) *olim*; left of the road from Frimley to Frimley Green (?); Pirbright Common, in damp bushy places; Holmwood, near Dorking, in a hollow halfway up Cold Harbour Lane; side of forked pond, Witley, next Thursley Common; Coleman Moor and Early Heath, Berks.

PILULARIA GLOBULIFERA.—Margins of lakes and ponds, not very frequent. Ponds on Putney Heath; Esher Common;* Mitcham Common;* Earlswood;* pond at Northaw, Herts; Worplesdon; Holmwood Common; Ditton Marsh; Hillingdon Common, Uxbridge.

POLYPODIUM PHEGOPTERIS.—Rare, Tilgate Forest.

POLYPODIUM VULGARE.—Rocks, walls, trunks of trees, banks, frequent. Hind Head Lane, near Godalming; road near Dorking; roadsides about Wendlesham; banks of the Wey, in a plantation N. of Frensham Common in profusion;* Keston Common; Croham Hurst;* between Edmonton and Winchmore Hill; about Cobham; Epping Forest,* &c.

POLYSTICHUM.—*See* ASPIDIUM.

PTERIS AQUILINA.—Heaths, woods and commons, frequent and abundant. Hampstead Heath;* Epping Forest;* &c.

SCOLOPENDRIUM VULGARE.—Banks and rocks in cold damp situations, not common. Merton Abbey walls; Chislehurst Church; N. side of Loughton Church; lanes about Chipping Ongar; Lee Churchyard; Epping Forest.

MINOR CRYPTOGAMS.

ALICULARIA.—*See* JUNGERMANNIA SCALARIS.

AMBLYSTEGIUM.—*See* HYPNUM.

ANACALYPTA LANCEOLATA.—Walls, &c., in calcareous districts. Near Dorking chalk-pits.*

ANDRÆA RUPESTRIS.—Damp rocks, very rare. High Rocks, Tunbridge Wells.

ANEURA.—*See* JUNGERMANNIA MULTIFIDA.

ANOMODON VITICULOSUS.—Trunks of trees, rare. In a lane near Kingsbury;* Southfield Park, Tunbridge Wells; Morant's Court Hill; about Northfleet and Gravesend; Reigate Hill.

ANTHOCEROS PUNCTATUS.—Ditches, &c., rare. Reported from Tunbridge Wells; Woolwich and Walthamstow; *olim.*

ANTITRICHIA CURTIPENDULA.—Damp rocks, very rare. Eridge Park; Ashdown Park.

ATRICHUM UNDULATUM.—Moist, shady banks and in woods, frequent. Epping Forest; and bordering banks and copses;* Wormley Wood,* &c.

AULOCOMNION ANDROGYNUM.—Banks, not frequent: by the roadside, in the hedge, leading from Tottenham to Walthamstow;* banks of an old gravel-pit in Epping Forest, between Walthamstow and Woodford;* sand rocks on Tunbridge Wells Common.

AULOCOMNION PALUSTRE.—Bogs, frequent. Epping Forest between Walthamstow and Woodford;* Hampstead Heath;* Harrow Weald Common; Stanmore Heath;* Barnes Common;* Surrey heaths.*

BARBULA MURALIS.—*See* TORTULA MURALIS.

BARTRAMIA FONTANA.—Swampy places, generally near springs, not unfrequent. Barnes Common, a patch or two;* Hampstead Heath, a patch or two in upper part of the bog;* Wimbledon Common, in one of the ravines, abundantly;* swampy field, Witley Park; swamps, Leith Hill; in a small pond on Shirley Common.

BARTRAMIA POMIFORMIS. Hedgebanks. Bank on the road to Haslemere from Godalming; bank S. end of Oxshott Hill;* bank on the road from Boreham Wood station to Stanmore Heath;* Leith Hill; Buckland Hill.

BRACHYTHECIUM.—*See* HYPNUM.

BRYUM PENDULUM.—A variety of *B. cæspitium*, from which it is distinguished by merely nominal differences. (Epping Forest.)

BRYUM CÆSPITIUM.—Walls, roofs, &c. (Common on almost every old wall about London.*)

BRYUM CAPILLARE.—In similar situations, common. (Old walls about London; everywhere.*)

BRYUM ERYTHROCARPUM.—Heaths, walls, &c. (Banks of the canal, Tottenham.*)

BRYUM ATRO-PURPUREUM.—Dry pastures, walls, &c. Hampstead Heath.*

BRYUM ARGENTEUM.—Old walls and banks, frequent. Old walls about London;* banks of the canal opposite Tottenham, plenty;* &c., Hampstead Heath.*

BRYUM PALLENS.—Near springs, not uncommon. Bogs, Putney Heath;* Epping Forest, between Woodford and Walthamstow;* Putney Heath.*

BRYUM TURBINATUM.—Moist places in sand and gravel-pits. Specimens from some locality near London;* in author's herbarium, but record wanting; Wormley Wood (?).

BRYUM ROSEUM.—Grassy banks and heaths, rare. Sandy banks and heaths in the Wealden, near Tunbridge Wells.

BRYUM LIGULATUM.—*See* MNIUM UNDULATUM.

BRYUM AFFINE.—*See* MNIUM AFFINE.

BRYUM CARNEUM.—*See* WEBERA CARNEA.

BRYUM ALBICANS.—*See* WEBERA ALBICANS.

BRYUM NUTANS.—*See* WEBERA NUTANS.

BRYUM PYRIFORME.—*See* LEPTOBRYUM PYRIFORME.

BRYUM, other than above.—*See* MNIUM.

CALYPOGEIA.—*See* JUNGERMANNIA TRICHOMANES.

CAMPYLOPUS TORFACEUS.—On peaty soils mostly, and where peat has been cut. Barnes Common; Hampstead Heath; Keston Common.

CATHARINEA.—*See* ATRICHUM.

CERATODON PURPUREUS.—Waste ground, banks and wells, common. Walls &c., about London; everywhere;* Hampstead Heath,* &c.

CHEILOSCYPHUS.—*See* JUNGERMANNIA POLYANTHOS.

CINCLIDOTUS FONTINALIS.—In rocks and streams, chiefly in calcareous districts (?).

CLIMACIUM DENDROIDES.—Meadows, bogs and marshes. Foot of Leith Hill; Nutfield Marsh.

CRYPHÆA HETEROMALLA.—Trunks of trees, rare. Epping Forest, near Loughton; in Enfield Chase *olim;* Kenton Lane, near Harrow Weald;* on a tree, foot of Box Hill.

CYLINDROTHECIUM CONCINNUM.—Rare, Box Hill.

DICRANELLA CERVICULATA.—Moist banks and heaths, especially where turf has been cut, frequent. Hampstead Heath;* Epping Forest; near High Beech; Barnes Common;* Surrey heaths.*

DICRANELLA VARIA.—Moist banks and clayey soils, (Harrow Weald Common).*

DICRANELLA CRISPA.—Moist sandy soil, near Southgate.

DICRANELLA HETEROMALLA.—Shaded banks. Epping Forest; Harrow Weald Common;* Hampstead Heath.*

DICRANUM SCOPARIUM.—Woods and copses, frequent. Epping Forest;* Hampstead Heath;* Pinner woods.*

DICRANUM MAJUS.—A variety of the above, with aggregate fruit-stalks; Pinner.

DIDYMODON RUBELLUS.—Wall-tops, common. Plentiful about Tottenham;* Edmonton;* Walthamstow;* on walls.

DIPHYSCIUM FOLIOSUM.—Moist rocks, rare. Sandstone rocks, Eridge; near Tunbridge.

EPHEMERUM SERRATUM.—On the bare ground, (W. side of Muswell Hill).

EUCALYPTA VULGARIS.—Walls and rocks, rare. Banks and wall-tops about Harefield (?); fissures of rocks in the Wealden about Tunbridge Wells and Eridge.

EURYNCHIUM.—*See* HYPNUM.

FEGATELLA CONICA.—Ditch banks, and by streams; rare.

FISSIDENS BRYOIDES.—Shady banks, not very common. Hole Lane, Kingsbury;* Harrow Weald Common;* Hampstead Heath.*

FISSIDENS ADIANTOIDES.—Shady places, wet rocks, banks, pastures and bogs, rare, Wormley Wood.*

FISSIDENS TAXIFOLIUS.—Moist clayey banks. Harrow Weald Common; lanes near* Hole Lane, Kingsbury.*

FONTINALIS ANTIPYRETICA.—On stones and on banks, both in running and in still water, common. Thames about Chertsey;* Walton,* &c.;

in the Lea and canal;* in a pond, by Kenton Lane, near Harrow Weald (fruiting).*

FOSSOMBRONIA.—*See* JUNGERMANNIA PUSILLA.

FRULLANIA.—*See* JUNGERMANNIA TAMARISCINA and DILATATA.

FUNARIA HYGROMETRICA.—On the ground, and on banks, common. Roadsides about London;* on Hampstead Heath;* waste places, West-end;* banks of the canals,* &c.

GRIMMIA PULVINATA.—Wall-tops, stones, &c.; common, about London on old mud-topped walls.* Wall, White Hart Lane, Tottenham;* banks of the canal there, near the lock this side of the rifle-butts.*

GRIMMIA APOCARPA.—Stones and wall-tops, rare. Colney Hatch Asylum wall;* sand rocks near Tunbridge Wells; Erith, &c.

GYMNOCIBE.—*See* AULOCOMNION.

GYMNOSTOMUM.—*See* POTTIA and PHYSCOMITRIUM.

HOMALIA TRICHOMANOIDES.—Trunks of trees, frequent. Lane near Tottenham Gasworks;* Epping Forest, below Woodford;* &c.

HOMALOTHECIUM SERICEUM.—Trees, stone walls, very common, scarce in the immediate neighbourhood of London. Not observed nearer than on a wall near the old church at Stanmore;* wall by the entrance to Bushy Park;* walls, foot of Box Hill;* &c.

HOOKERIA LUCENS.—Moist banks, very rare. High Rocks, Tunbridge Wells; Erith Park.

HYLOCOMIUM.—*See* HYPNUM.

HYPNUM LUTESCENS.—Calcareous rocks and borders of wood; old quarries &c., near the sea. About Northfleet and Gravesend; on chalk roadside from Tunbridge Wells to Southborough; Reigate Hill.

HYPNUM NITENS.—Marshy and boggy ground in peat bogs, probably not unfrequent.

HYPNUM GLAREOSUM.—Gravelly, grassy banks, not uncommon. Roadside from Woodford to High Beech, Epping Forest; in a gravel-pit behind Tottenham;* Harrow.

HYPNUM ALBICANS.—Dry grassy places, not uncommon. Fruit rare; Stanmore Heath;* Hampstead Heath; sandy places, Tunbridge Wells; Reigate Heath.

HYPNUM VELUTINUM.—On trunks of trees, stones, and on the ground, common. Everywhere in lanes &c., about London; Tottenham;* Hampstead Heath;* Chingford;* &c.

HYPNUM RUTABULUM.—In similar situations, very common, everywhere in the same localities as the preceding species,* (banks of the Lea Canal, &c.)

HYPNUM POPULEUM.—On stones and trees, frequent, but rarely fruiting about London. On a wall at Tottenham;* tombstones, Chingford Old Church;* on stones on the hill above Westerham.

HYPNUM PLUMOSUM.—On stones &c., in wet places, rare (?). By the Lea canal;* and by a drain adjoining,* on trees in Enfield Chase, *olim.*

HYPNUM ILLECEBRUM.—On grassy banks and rocky pastures, generally near the sea, rare. Sevenoaks; Barnes Common (Quekett Micr. Club report); hedgebanks near Shiere.

HYPNUM CRASSINERVIUM.—On limestone in shady places, rare. Wood, Box Hill; on the ground, and at the foot of trees.

HYPNUM PILIFERUM.—Shady banks and woods, or on stones, not uncommon, but rarely fruiting. Putney Heath, in the hollow left of the Kingston road;* bank, roadside from Woodford to High Beech, Epping Forest.*

HYPNUM PRÆLONGUM.—Moist shady banks, common. Epping Forest;* and outlying copses;* Wormley Wood;* general in woods,* Hampstead Heath.* Var. *Schwartzii* on hedge-banks near Shiere.

HYPNUM PUMILUM.—Hedgebanks, sandstone rocks, rare. Wood, Box Hill; on the ground, and at the foot of trees.

HYPNUM MYOSCUROIDES.—Rare. Sandstone rocks about Tunbridge Wells; Harrow Weald Common; Gatton woods.

HYPNUM STRIATUM.—On the ground in woods, not unfrequent. Chingford Hatch;* Epping Forest;* Stroud Copse, Godalming; Pinner Wood.*

HYPNUM RUSCIFOLIUM.—On stones in rivulets and sluices, frequent. Millstream, Chingford Mill;* canal banks, Tottenham;* Harrow; in the Mole, foot of Box Hill.

HYPNUM MURALE.—On rocks and stones, not frequent. On the brickwork, mouth of a drain, lower part of Hampstead Heath;* on stones by the Tottenham Canal.*

HYPNUM CONFERTUM.—On stones, shady walls, trunks of trees, frequent. Foot of the brick wall between Highgate and Hampstead;* gravestones Walthamstow;* banks of the Tottenham Canal;* old wall on the Frant road, Tunbridge Wells; foot of trees, Box Hill; tree variety, in a lane near Stanmore Heath.

HYPNUM MEGAPOLITANUM.—Sandy banks, (var. of the preceding according to Hooker and Wilson). Vicinity of Dorking; Gomshall and Shiere, ('Science Gossip,' February, 1872).

HYPNUM TENELLUM.—Walls and rocks, especially such as are calcareous, common. On tombstones, Chingford Old Church;* foot of trees, Box Hill; wall in a lane near Tottenham Church.*

HYPNUM SERPENS.—Moist banks, trunks of trees, and on stones near water, common. On stones by the canal about Tottenham;* tombstones in Kensal Green Cemetery;* &c.

HYPNUM IRRIGUUM.—On stones in rivulets, mill-dams, &c., rare. About Dorking; Gomshall and Shiere; a variety of the above (?).

HYPNUM RIPARIUM.—Banks of streams and moist places on the ground, common, (plentiful on the banks of the Lea Canal*).

HYPNUM MEDIUM.—*See* LESKEA POLYCARPA, &c.

HYPNUM STELLATUM.—Boggy places, not common. Bog on Winter Downs, Esher; (slope east),* a patch on Putney Heath, near entrance to Roehampton Lane;* on Bagshot Heath, under the hills left of Bagshot;* Tunbridge Wells Common; Leith Hill; (Bagshot locality in prostrate form).

HYPNUM CHRYSOPHYLLUM.—Marsh lands, and on fallow-fields, rare, a variety of the above (?), reported as occurring on Hampstead Heath; Harrow Weald Common (?); at the foot of Buckland Hill.

HYPNUM PALUSTRE.—On stones in rivulets of subalpine countries, rare. Hampstead Heath (?); watery places, and on rotten wood about Tunbridge Wells; below Red Hill.

HYPNUM STRAMINEUM.—Bogs, rare. Hampstead Heath;* scarce, in one place, upper part of the bog; about Dorking; bogs in Ashdown Forest; Reigate Heath; Gomshall and Shiere ('Science Gossip'); perhaps a slender and fruiting form of the following:—

HYPNUM CORDIFOLIUM.—Marshy places and ditches, not unfrequent. Epping Forest; between Woodford and Walthamstow;* copse borders of Harrow Weald Common;* Pinner, not fruiting.

HYPNUM CUSPIDATUM.—Bogs and marshes, common. Barnes Common;* Putney Heath;* Hampstead Heath;* Epping Forest; in pools by the road to Woodford in fine condition.*

HYPNUM SCHREBERI.—In woods among trees and bushes, not very common, unless passed over in mistake for the following species. Epping Forest;* Pinner Woods; on heaths near Tunbridge Wells; and at High Rocks; in fruiting condition.

HYPNUM PURUM.—Shady banks, common, but rarely fruiting, (Epping Forest;* &c.)

HYPNUM SPLENDENS.—On the ground in woods, common. Epping Forest;* Stanmore Heath;* Burnham Beeches.*

HYPNUM BREVIROSTRE.—In mountainous woods, rare, near Tunbridge Wells.

HYPNUM IMPONENS.—Box Hill ('Science Gossip').

HYPNUM SQUARROSUM.—Wet pastures, and in woods, common, rare in fruit. Epping Forest;* Harrow Weald;* &c.

HYPNUM TRIQUETRUM.—Woods and banks, common. Epping Forest;* Broxbourne and Wormley woods;* Box Hill, &c.*

HYPNUM LOREUM.—Woods in hilly countries, not very frequent. Epping Forest, near Chingford Hatch;* Darent Wood;* Box Hill, &c.;* Shiere; Harrow Weald Common;* Eridge Rocks.

HYPNUM ADUNCUM.—Bogs, not very common. Boggy pits in Epping Forest;* in a small pond on Shirley Common; bogs near Tunbridge Wells; Reigate Heath.

HYPNUM FLUITANS.—Meadows, marshes, and peat bogs, common. Hampstead Heath;* ditches by the railway above Lea Bridge station, in perfection, and fruiting freely.*

HYPNUM UNCINATUM and var. *H. exannulatum*.—In watery stony places, in hilly countries. Harrow Weald Common; in the vicinity of Dorking, Gomshall, and Shiere.

HYPNUM FILICINUM.—Watery places, generally in hilly countries, rare. Hole Lane, near Kingsbury;* about Northfleet and Gravesend; Chislehurst; about Tunbridge Wells.

HYPNUM COMMUTATUM.—Possibly about rills in the chalk districts.

HYPNUM MOLLUSCUM.—Hilly places in the chalk districts, not unfrequent. Box Hill;* Mickleham;* Reigate Heath (a variety?); Morant's Court Hill; Reigate Hill, rarely fruiting.

HYPNUM CUPRESSIFORME.—On banks, stones, trunks of trees, common, and very variable in appearance; everywhere, especially in woodland districts. Hampstead Heath;* also a pale-leaved variety;* Epping Forest;* Darent Wood * (varieties), &c.

HYPNUM DENTICULATUM.—In woods and on banks, common. Hedgebanks about London, frequent; Tottenham;* Kingsbury;* Chingford, &c.;* Reigate Heath.

HYPNUM SYLVATICUM.—Roots of trees in woods, not common. Epping Forest;* Darent Wood;* Harrow Weald Common.*

HYPNUM UNDULATUM.—Woods, and heathy places, in hilly parts, rare. Eridge Park; lane W. of Reigate Park?

ISOTHECIUM MYURUM.—On walls and on trees in woods, not common. Leith Hill; trees, &c., about Tunbridge Wells.

ISOTHECIUM MYOSUROIDES.—*See* HYPNUM.

JUNGERMANNIA ASPLENIOIDES.—Shady banks and moist woods, frequent, but rarely fruiting. Copse near Chingford Hatch, and elsewhere in Epping Forest;* Erith Rocks (fruiting in 1839); Pinner Wood; Wormley Wood;* Leith Hill.[1] (Plagiochila.)

JUNGERMANNIA CURVIFOLIA.—Reigate Heath. (Plagiochila.)

JUNGERMANNIA BISCUSPIDATA.—Trees, hedgebanks, and moors, frequent. Hampstead Heath;* Epping Forest;* Pinner Wood; Barnes Common; Leith Hill.

JUNGERMANNIA MULTIFIDA.—Wet places, frequent. Hampstead Heath;* Hole Lane.* (Aneura;* Riccardia.)

JUNGERMANNIA ALBICANS.—Woods and banks, common. Roadside, &c., from Hampstead to Highgate;* Reigate Heath;* Epping Forest.*

JUNGERMANNIA EPIPHYLLA.—Sand-rocks about Tunbridge Wells, &c., frequent. Woolwich; gutters in Dean's Wood, Harefield; Leith Hill; Barnes Common. (Pellia; Metzgeria.)

JUNGERMANNIA CALYCINA.—Boggy and wet places. (Pellia.)

JUNGERMANNIA FURCATA.—Trunks of trees, common. Moors, &c., Epping Forest, borders; Leith Hill.

JUNGERMANNIA CRENULATA.—Heaths, &c., common. Hampstead Heath;* Harrow Weald Common.*

JUNGERMANNIA TOMENTOSA.—Wood between Highgate and Hornsey. (Trichocolea.)

JUNGERMANNIA SPHÆROCARPA.—Old Fall Wood, Highgate; by a rill.

JUNGERMANNIA CILIARIS.—In Old Fall Wood, Highgate; by a rill. (Ptylidium.)

JUNGERMANNIA TAMARISCINA.—Chalk districts on the ground, common. Hampstead. (Frullania.)

JUNGERMANNIA UNDULATA.—Harrow Weald Common;* Hampstead Heath;* Leith Hill. (Scapania.)

[1] See remarks in the Preface. For practical purposes the student may arrange his collection in two principal groups, Foliaceous and Frondose; the former subdivided according to the presence or absence of stipules (so called) on the stems. The latter form a natural transition to the Marchantiaceæ.* See M. C. Cooke, 'British Hepaticæ.' R. Hardwicke and Bogue. Price 4*d*.

JUNGERMANNIA COMPLANATA.—Trunks of trees, frequent. Hedgebanks about Chingford Hatch;* Pinner Wood;* Hole Lane, Kingsbury;* Leith Hill; Barnes Common. (Radula; Martinellia.)*

JUNGERMANNIA HETEROPHYLLA.—With *J. biscuspidata* on Barnes Common and Hampstead Heath. (Lophocolea.)

JUNGERMANNIA BIDENTATA.—Boggy places, frequent. Pinner Wood; Hampstead Heath;* Hole Lane;* Barnes Common; Leith Hill. (Lophocolea.)

JUNGERMANNIA TRICHOMANES.—Sand-rocks, bogs, &c. Epping Forest;* Pinner Wood. (Calypogeia.)

JUNGERMANNIA POLYANTHOS.—Wet places about streams. Epping Forest.* (Cheiloscyphus.)

JUNGERMANNIA REPTANS.—Hedgebanks, heaths, and woods, common. Epping Forest;* Hampstead Heath;* Leith Hill. (Lepidozia.)

JUNGERMANNIA PLATYPHYLLA.—Walls, rocks, and trees, common. On trees about Chingford Hatch;* and in the forest;* Reigate Heath.* (Madotheca.)

JUNGERMANNIA DILATATA.—Trunks of trees, common. Pinner Wood. (Frullania.)

JUNGERMANNIA INFLATA.—Frequent. Hampstead Heath;* Reigate Heath.*

JUNGERMANNIA SPHAGNI.—Bogs. (Physiolium.)

JUNGERMANNIA PUSILLA.—Moist shady banks, rare; about Woolwich. (Fossombronia.)

JUNGERMANNIA SCALARIS.—Boggy places. Hampstead Heath;* Harrow Weald Common;* Leith Hill. (Alicularia.)

JUNGERMANNIA ANOMALA.—On trees,* no record.

JUNGERMANNIA CONNIVENS.—Bogs. Hampstead Heath.*

JUNGERMANNIA VARIA.—Woods near Harefield.

JUNGERMANNIA DIVARICATA.—Hampstead Heath;* Reigate Heath.

JUNGERMANNIA EMARGINATA.—Leith Hill. (Sarcoscyphus.)

JUNGERMANNIA NEMOROSA.—Woods in the Wealden, abundant. (Scapania.) —N.B. Many other Jungermanniæ are reported to occur in the neighbourhood of Tunbridge Wells; High Rocks; Eridge Rocks, &c.

LEPIDOZIA.—*See* JUNGERMANNIA REPTANS.

LEPTOBRYUM PYRIFORME.—Sandy or turfy ground, sandstone rocks; in stoves and greenhouse frames, sometimes; rare. Barnes Common; sand-rocks, Tunbridge Wells; Dorking; Shiere; Gomshall Marsh.

LEPTODON SMITHII.—On trunks of trees, rare. In the vicinity of Dorking, Gomshall, and Shiere.

LEPTOTRICHUM FLEXICAULE.—On calcareous rocks, rare. Slopes of Box Hill.

LESKEA POLYCARPA.—At the roots of trees, on stones and piles near water. Harrow Weald; trunks of Willows, Chiswick; on trees by the Brent at Stonebridge Harrow Road;* ditch by the Thames on stumps of Willows, near Kew.*

LESKEA SERICEA.—*See* HOMALOTHECIUM.

LEUCOBRYUM GLAUCUM.—Damp places in upland woods, not common.

Epping Forest;* Harrow Weald Common;* Esher Common (on the ridge, under the Pines);* Burnham Beeches;* forests about Tunbridge Wells; nowhere fruiting.

LEUCODON SCIUROIDES.—Trunks of trees, rare. Harrow Weald Common; forests about Tunbridge Wells; copses on the chalk range in several places,* rarely fruiting.

LOPHOCOLEA.—*See* JUNGERMANNIA BICUSPIDATA, BIDENTATA, HETEROPHYLLA.

MADOTHECA.—*See* JUNGERMANNIA* PLATYPHYLLA.

MARCHANTIA POLYMORPHA.—Wet places, bogs, sides of drains, &c., not unfrequent. On a wall by the rivulet at Edmonton, Angel Lane;* Hampstead Heath, under wet banks;* Wimbledon Common, in one of the ravines, plenty.* Banks of holes, and old gravel-pits in Epping Forest, between Woodford and Walthamstow.*

MARCHANTIA CONICA.—*See* FEGATELLA.

MARTINELLIA.—*See* JUNGERMANNIA COMPLANATA.

METZGERIA.—*See* JUNGERMANNIA EPIPHYLLA, FURCATA.

MNIUM UNDULATUM.—Woods and shady banks, not uncommon; but rare in fruit. Harrow and Pinner woods; Epping Forest;* Hole Lane, Kingsbury;* copse by the forked ponds, Witley Common;* E. Grinstead, fruiting; Leith Hill; Buckland Hill.

MNIUM AFFINE.—Woods and marshes, rare. High Rocks, Tunbridge Wells.

MNIUM ROSTRATUM.—Shady spots, near springs, rare. Harrow Weald Common; Gomshall Marsh.

MNIUM HORNUM.—Woods, shady banks, common. Epping Forest;* Harrow Weald Common;* Burnham Beeches,* &c.; Hampstead Heath.

MNIUM PUNCTATUM.—Shady banks near springs, &c., not uncommon. Stroud Copse, Godalming; Leith Hill;* Epping Forest;* Wormley Wood;* Hampstead Heath (a patch of it in a ditch).*

NECKERA COMPLANATA.—On trees in woods, &c., not very frequent. Epping Forest;* Harrow Weald, neighbouring lanes,* fruit rare.

NECKERA CRISPA.—On trunks of trees, and on calcareous rocks, rare. Morant's Court Hill; about Dorking, Gomshall, and Shiere; chalk cliffs, Gravesend; Epsom Downs (?); Buckland and Reigate Hills.

NECKERA PUMILA.—In similar situations, rare. About Dorking, Gomshall, and Shiere; near Tunbridge Wells, road to Frant.

ORTHOTRICHUM CUPULATUM.—On stones, trunks of trees especially in calcareous districts, rare. (?) Darent Wood.*

ORTHOTRICHUM ANOMALUM.—On rocks and walls, especially in calcareous countries about Dorking, Gomshall, and Shiere (Godalming, on a roof).

ORTHOTRICHUM TENELLUM.—On trees, rare. About Dorking, Gomshall, and Shiere.

ORTHOTRICHUM AFFINE.—On trees and stones, common. Epping Forest;*

walls of old Chingford Church ;* Pinner ; Kenton Lane, near Harrow Weald.*

ORTHOTRICHUM STRAMINEUM.—Trees and boulders, rare. Vicinity o Dorking, Gomshall, and Shiere.

ORTHOTRICHUM BRUCHII.—*See* ULOTA BRUCHII.

ORTHOTRICHUM DIAPHANUM.—On trees and stones, common. Epping Forest.

ORTHOTRICHUM PULCHELLUM.—On trunks of trees, and stones, rare. (?) Vicinity of Dorking, Gomshall, and Shiere.

ORTHOTRICHUM STRIATUM.—*See* O. LEIOCARPUM.

ORTHOTRICHUM LEIOCARPUM.—On trees, pales, and stones. On a tree at Stroud ; neighbourhood of Box Hill and Dorking.

ORTHOTRICHUM LYELLII.—On trunks of trees, rare. Vicinity of Box Hill and Dorking.

ORTHOTRICHUM CRISPUM.—*See* ULOTA CRISPA.

PELLIA.—*See* JUNGERMANNIA EPIPHYLLA and J. CALYCINA.

PHASCUM CUSPIDATUM.—On the ground, especially in sandy soil, common. On a bank right, Chingford Mills ; road to Chingford.*

PHASCUM BRYOIDES.—Open fields, rare. * No record where gathered.

PHASCUM MUTICUM.—*See* SPHÆRANGIUM.

PHASCUM SUBULATUM.—*See* PLEURIDIUM SUBULATUM.

PHASCUM SERRATUM.—*See* EPHEMERUM.

PHILONOTIS.—*See* BARTRAMIA FONTANA.

PHYSCOMITRIUM PYRIFORME.—Moist banks, common. Frequent by the Lea Canal ;* Gomshall Marsh.

PHYSIOTIUM.—*See* JUNGERMANNIA SPHAGNI.

PLAGIOCHILA.—*See* JUNGERMANNIA ASPLENIOIDES and J. CURVIFOLIA.

PLAGIOTHECIUM.—*See* HYPNUM PULCHELLUM and H. DENTICULATUM.

PLEURIDIUM SUBULATUM.—On the ground in fields, and on banks, common. Hampstead Heath ;* Epping Forest ;* Barnes Common.

PLEUROZIUM.—*See* HYPNUM.

POGONATUM NANUM.—Heaths, sandy or loamy banks, not uncommon. Surrey heaths in several places ;* Epping Forest.*

POGONATUM ALOIDES.—Heaths, moist banks, frequent. Barnes Common ; Pinner ; Harrow Weald Common.*

POLYTRICHUM PILIFERUM.—On dry open heaths, common. Hampstead Heath.*

POLYTRICHUM JUNIPERINUM.—On heaths, frequent. Barnes Common.*

POLYTRICHUM COMMUNE.—Moist woods, and boggy parts of heaths, common. Hampstead Heath ;* Putney Heath ;* Barnes Common,* &c.

POTTIA CAVIFOLIA (*olim* GYMNOSTOMUM OVATUM).—Clay walls, and the naked ground, common.

POTTIA TRUNCATA.—Banks and fallow ground, frequent. Epping Forest ;* Hampstead Heath.*

PSEUDOLESKEA CATENULATA.—Sub-alpine rocks, rare. High Rocks, Tunbridge Wells.

PSYCHOMITRIUM.—*See* LEPTOBRYUM.

PTERYGONIUM GRACILE.—Rocks, walls, and trunks of trees, in sub-alpine districts, rare. In the vicinity of Dorking, Gomshall, and Shiere; Ramsbye Rocks, near Tunbridge.

PTERYGONIUM SMITHII.—*See* LEPTODON SMITHII.

PTILIDIUM.—*See* JUNGERMANNIA CILIARIS.

RACOMITRIUM CANESCENS.—Sandy ground on heaths, &c.; rare near London. Enfield Chase (*olim*); Rusthall Common; near Croydon (?) *olim*; Leith Hill.

RACOMITRIUM LANUGINOSUM.—Walls and rocks, &c., dry woods, and sandy hilly places. Near Tunbridge Wells, rare.

RACOMITRIUM HETEROSTICHUM.—Rocks and walls, in dry places, rare. High Rocks, Tunbridge Wells.

RADULA.—*See* JUNGERMANNIA COMPLANATA.

RHYNCOSTEGIUM.—*See* HYPNUM.

RICCARDIA.—*See* JUNGERMANNIA MULTIFIDA.

RICCIA CRYSTALLINA.—On banks, &c., probably common, but overlooked.

SARCOSCYPHUS.—*See* JUNGERMANNIA EMARGINATA.

SCAPANIA.—*See* JUNGERMANNIA NEMOROSA and UNDULATA.

SCHISTOSTEGA OSMUNDACEA.—Moist banks and sandstone caves, rare. In the vicinity of Dorking, Gomshall, and Shiere.

SCHISTOSTEGA PINNATA.—*See* preceding.

SCLEROPODIUM.—*See* HYPNUM ILLECEBRUM.

SELIGERIA CALCAREA.—Steep sides of chalk-pits, local. In the vicinity of Dorking, Gomshall, and Shiere.

SPHÆRANGIUM MUTICUM.—Fallow fields, &c., not uncommon.

SPHAGNUM ACUTIFOLIUM.—Bogs and swamps, frequent. Hampstead Heath;* Esher Common; slope E. of Winter Downs.*

SPHAGNUM CUSPIDATUM.—Deep peat bogs. A variety of the above, in similar situations.

SPHAGNUM CYMBIFOLIUM.—Bogs and swamps, the usual form. Everywhere in bogs; Hampstead Heath,* Surrey heaths, &c.*

SPHAGNUM SQUARROSUM.—A mere variety; growing where there is abundance of water. There is no *squarrose* variety of *cymbifolium*, &c., different to this, except in the imagination of species splitters; other varieties recorded in the books.

SPLACHNUM AMPULLACEUM.—On the dung of herbivorous animals, rare. High part of Ashdown Forest, not far from Wych Cross, plentiful; great bog near Forest Row; Leith Hill.

STEREODON.—*See* HYPNUM CUPRESSIFORME.

TETRAPHIS PELLUCIDA.—On the ground; generally on peat banks, rare. Staines; Ken Wood, by sides of carriage drive; sandstone rocks about Tunbridge Wells, abundant; in the vicinity of Dorking, Gomshall, and Shiere.

TETRAPHIS BROWNIANUM.—*See* TETRODONTIUM.

TETRODONTUM BROWNIANUM.—On gritty, sandy rocks, rare. Eridge Rocks, near Tunbridge Wells.

THAMNIUM ALOPECURUM.—Moist, shady woods, and on rocks, rare. Leith Hill; Harrow Weald (?) *olim*; wood, Box Hill.*

THUIDIUM TAMARISCINUM.—Woods and banks, common. Epping Forest,* Stanmore Heath,* Box Hill,* Darent Wood,* &c.

THUIDIUM HYSTRICOSUM (a variety of *tamariscinum*). Foot of Buckland Hill.

THUIDIUM ABIETINUM.—Dry banks in the chalk districts, rare. Slopes of Box Hill; Hog's Back; Morant's Court Hill; about Northfleet and Gravesend.

THUIDIUM DELICATULUM (a var. of *tamariscinum*). In the vicinity of Dorking, Gomshall and Shiere.

TORTULA RURALIS.—Walls, banks, and especially on thatched roofs, common. At Cricklewood;* Dorking; in almost every country village or hamlet.*

TORTULA SUBULATA.—On banks, and about the roots of trees, frequent. Hedgebank near Stroud; Pinner;* Harrow Weald;* by the Mole, near Esher Mills;* roadside from Woking station to Guildford,* &c.

TORTULA MURALIS.—Walls, common. Everywhere about London; Hampstead,* Tottenham,* Hammersmith,* &c.

TORTULA MARGINATA.—On walls and on the ground, rare. In the vicinity of Dorking, Gomshall, and Shiere.

TORTULA LATIFOLIA.—On trees, rare. In the vicinity of Dorking, Gomshall, and Shiere; Epping Forest* (?).

TORTULA CUNEIFOLIA.—Banks, rare (?). In the vicinity of Dorking, Gomshall, and Shiere.

TORTULA UNGUICULATA.—On the ground, frequent, and on banks by the roadside. Harrow Weald Common;* near Colney Hatch.*

TORTULA ALOIDES.—On clay banks, not uncommon, but rare; near London;* but no record of where found.

TORTULA HORNSCHUCHIANA.—Walls, rocks, and on the ground. Near Dorking, Shiere, &c.; Laleham.

TORTULA VINEALIS.—On walls. About Dorking, Gomshall, and Shiere.

TORTULA FALLAX.—Walls, fields, stony ground, common. Ruislip; Harrow Weald Common.*

TRICHOCOLEA.—*See* JUNGERMANNIA TOMENTOSA.

TRICHOSTOMUM CRISPULUM.—On limestone rocks near the sea. In the vicinity of Dorking, Gomshall, and Shiere.

ULOTA BRUCHII.—On trees, rare. In the vicinity of Dorking, Gomshall, and Shiere.

ULOTA CRISPA.—On trees, common. Pinner Woods; Stroud Copse, Godalming.

WEBERA NUTANS.—Wet places on heaths, common. Hampstead Heath;* Epping Forest.*

WEBERA CARNEA.—Moist, clayey banks, rare (?). Near Reigate;* Blackheath; between Lee and Eltham; about Dorking.*

WEBERA ALBICANS.—Moist, sandy ground, not uncommon. Putney Heath;* Gomshall; Hampstead Heath;* Stanmore Heath.*

WEISSIA CONTROVERSA.—Pastures and fallow fields, frequent. Epping Forest;* and on a bank by the road to Chingford Mill, from Edmonton.*
WEISSIA CIRRHATA.—On old railings and thatched roofs, frequent. In Roehampton Lane;* roof at Cricklewood;* Harrow, road to Harrow Weald.*

ZYGODON VIRIDISSIMUS.—Trunks of trees, rare (?). Darent Wood;* woods about Tunbridge Wells.

MINOR CRYPTOGAMS: THALLOGENS,[1]
LICHENS.

LICHENS.—Rare in the immediate neighbourhood of London, but plentiful in Epping Forest; Wormley and Broxbourne woods; Burnham Beeches and other woodland parts; Harrow Weald Common; Surrey heaths, &c.; old stones near churchyards, &c.; old walls.
The following are but a small portion of what may be found by collectors:

ARTHONIA ASTROIDEA.—Trunks of trees.

BOOMYCES RUFUS.—Rare, heaths.

CALYCIUM CURTUM, TRICHIALE.—Old trees and planks.
CLADONIA COCCIFERA, CORNUTUS, FIMBRIATA, FURCATA, PYXIDATA, RADIATA, RANGIFERA. SYLVATICA.—Heaths. Surrey, and Bucks; Keston Common;* Harrow Weald Common;* Pinner Wood.

EVERINA PRUNASTRI.—Old trees and woods.

GRAPHIS STRICTA.—Trunks of trees.

LECIDIA.—Several species. Walls, heaths, bark of trees.
LECONORA SUBFUSCA and varieties.—Trees, old walls, &c. L. VITELLINA, old palings.
LEPRARIA VIRIDIS, FLAVA, ALBA.—Trunks of trees.

OPEGRAPHIS ATRA, BETULINA.—Bark of trees.

[1] The study of the minor Cryptogams, of this section of them especially, is difficult, owing to the chaos of nomenclature and terminology which obtains: a partial acquaintance with the subject is, however, indispensable to the student. Most of the typical forms of British Lichens may be learned from the species indicated, and specimens of Fungi should be collected which may illustrate the various ways in which the spores are arranged; as, for instance:—
1. Spores naked:—on plates, Agaricus; on spines, Hydnum; in pores, Boletus; &c.
2. Spores free, enclosed in a sac:—Lycoperdon. Other forms, Geaster, Phallus, &c.
3. Spores in spore-cases, exposed or immersed in the substance of the plant:—Helvella, Tuber, &c.

As for the vast and formidable array of anomalous vegetable growths known as blights, mildews, mould, oak-spangles, etc., they have no claim to be classed as fungals, whatever may be said for Clavaria, Sphæria, Peziza, and other amorphous vegetations of a membranous, horny, fleshy, and even gelatinous structure. For instructions as to the best method of preserving these minor Cryptogams, see 'Science Gossip,' September and October, 1872.

PARMELIA.—Several species on trunks of trees, palings, &c. P. SAXATILIS, stones. P. PARIETINA, walls. P. PHYSODES, heaths.

PELTIGERA CANINA.—On mossy banks. P. POLYDACTYLA, moist places in shady woods. P. RUFUS, woods and banks.

PERTUSSARIA COMMUNIS, FALLAX.—Trunks of trees.

PHYSIA STELLATA, PARIETINA, CITRINELLA.—Old palings, roofs of houses, &c.

PLACODIUM MURORUM.—Gravestones.

PSORA OSTREATA.—On old palings.

RAMALINA CALICARIS, POLYMORPHA, and varieties.—On trees and hedges.

STIGNALIDIUM CRASSUM. Bark of trees.

SCYPHOPHORUS PYX.—*See* CLADONIA (?).

SPHÆRIA DISCIFORMIS.—Hampstead Heath.

SPILOMA MURALE.—Walls.

SQUAMARIA MURORUM.—Walls.

STICTA PULMONARIA.—Woods.

VARIOLARIA VITILIGO, DISCOIDEA, &c. Old posts and pales.

VERRUCARIA GEMMATA.—Trunks of trees. V. MURALIS, V. MUTABILIS, walls &c.

FUNGI.

Most of the species detailed below are important, either for their edible or for their poisonous qualities. A full list of what may be found within a ten-mile radius north of London is given, on the authority of Mr. Worthington Smith, by Messrs. Trimen and Dyer, in an appendix to their 'Flora of Middlesex.'

AGARICUS ÆRUGINOSUS.—Stumps of trees (Hampstead Heath).

AGARICUS ARVENSIS.—Pastures.

AGARICUS AURANTIUS.—Epping Forest.

AGARICUS BULBOSUS.—Epping Forest.

AGARICUS CAMPANULATUS.—Epping Forest.

AGARICUS CAMPESTRIS.—Pastures.

AGARICUS CARNOSUS.—Hampstead Heath.

AGARICUS CRUSTULINIFORMIS.—Woods.

AGARICUS DEALBATUS.—About Fir plantations.

AGARICUS EQUESTRIS.

AGARICUS FASCICULARIS.—Old stumps of trees.

AGARICUS FLOCCOSUS.—Charlton Wood.

AGARICUS FRAGILIS.—About New River.

AGARICUS GAMBOSUS.—Lanes and pastures.

AGARICUS GLUTINOSUS.—About New River.

AGARICUS INTEGER.

AGARICUS LACTIFERUS.—Charlton Wood.

Agaricus lateralis.—Epping Forest.
Agaricus muscarius.—Birch woods; Highgate Woods.
Agaricus nebularis.—Rare. Borders of woods.
Agaricus ostreatus.—Trunks of Elms, &c.
Agaricus personatus.—Rare, (pastures near Highbury, *olim*).
Agaricus phalloides.—Woods.
Agaricus piperatus.—Charlton Wood.
Agaricus prunulus.—Woods (N. of London).
Agaricus rubescens.—Woods.
Agaricus separatus.
Agaricus sinuatus.—Woods (N. of London).
Agaricus semiglobatus.—Pastures, &c.
Agaricus silvicola.—Woods (Highgate).
Agaricus sublateritius.—Woods, on stumps.
Agaricus sulfureus.—Shady places (S. of London).
Agaricus umbelliferus.—Epping Forest.
Agaricus vaginatus.—Pastures.
Agaricus vernus.—Woods, rare.
Agaricus viscosus.—Hampstead Heath.
Agaricus violaceus.—Woods.
Agaricus viscidus.—About New River.

BOLETUS ÆSTIVALIS.—Woods (Highgate).
Boletus bovinus.—Highgate woods.
Boletus edulis.—Woods.
Boletus felleus.—Woods (Epping Forest).
Boletus ignarius.—Woods (Wimbledon), and on Willows.
Boletus luridus.—Woods, common.
Boletus piperatus.—Woods, rare.
Boletus squamosus.—Hackney.
Boletus suberosus.—Clapton.
Boletus subtomentosus.—Woods (Esher).

CANTHARELLUS CIBARIUS.—Woods (Epping Forest).
Clavaria Hypoxylon.—Woods (Wimbledon), and on Willows.
Clavaria rugosa.—Woods, common.
Clavaria vermiculata.—Waysides and pastures.
Coprinus atramentarius.—Old stumps, &c.
Coprinus comatus.—Pastures (in the parks).
Coprinus picaceus.—Rare.
Cortinarius violaceus.—Open places in woods.

FISTULINA HEPATICA.—Trunks of Oak-trees in woods.
Fungus campaniformis.—Chipping Ongar.

HELVELLA CRISPA.—Shady places; rare near London.
Hydnum imbricatum.—About the Ravensbourne; wood near Tottenham.
Hydnum repandum.—Woods (N. of London).
Hygrophorus conicus.—Pastures and roadsides.

HYGROPHORUS VIRGINEUS.—Lawns and pastures.
HYGROPHORUS PSITTACINUS.—Rich pastures.

LACTUCARIUS ACRIS.—Woods.
LACTUCARIUS DELICIOSUS.—Fir plantations.
LACTUCARIUS PIPERATUS.—Woods, common.
LACTUCARIUS PYROGALUS.—Woods and meadows.
LACTUCARIUS RUFUS.—Fir woods.
LACTUCARIUS THEIOGALUS.—Woods (Hampstead).
LACTUCARIUS TORMINOSUS.—Woods, &c., not common.
LYCOPERDON GIGANTEUM.—Meadows (near Hampstead and Highgate).
LYCOPERDON PEDUNCULATUM (Hackney).
LYCOPERDON STELLATUM (Hackney).

MARASMIUS OREADES.—Pastures and roadsides.
MARASMIUS URENS.—In similar situations.
MORCHELLA ESCULENTA.—Rare.

PANUS STYPTICUS.—Old trees and stumps. In woods.
PEZIZA CORNUCOPOIDES.—About New River.
PEZIZA CYATHOIDES.
PEZIZA LENTIFERA.—Hackney.
PEZIZA PUNCTATA.
PHALLUS ESCULENTUS.—Wood near Dartford.
PHALLUS IMPUDICUS.—Woods, common (Hackney).

RUSSULA ALUTACEA.—Woods.
RUSSULA EMETICA.—Woods. In damp places.
RUSSULA FŒTENS.—Common.
RUSSULA HETEROPHYLLA.—Woods.
RUSSULA SANGUINEA.—Woods, frequent.

TREMELLA GRANULATA.—Hackney?
TREMELLA PURPUREA.—Hackney?
TUBER ÆSTIVUM.—Under trees.

UREDO EFFUSA.—Hampstead Heath.

ALGÆ.

ULVA COMPRESSA, INCRASSATA.
ULVA CRISPA.—Under walls.
ULVA LACTUCA.—Ditches.

At Southend other Algæ may probably be found washed up by the tides: Fucus vesiculosus, serratus, palmatus; Laminaria digitata; Plocamium coccineum, &c.

NOTES, ON THE AUTHORITY OF THE 'CYBELE BRITANNICA,' AND COMPENDIUM THERETO.

ADONIS AUTUMNALIS; imperfectly established.
Alisma natans; localities need verification.
Alopecurus bulbosus; a state, perhaps, of *A. geniculatus*.
Asparagus officinalis; often only a garden waif.
Asperugo procumbens; misreported, or extinct in many places.
Atriplex arenaria and *Babingtonia*; confused, formerly, with each other.
Atriplex erecta; scarcely separable from *A. angustifolia*.

BARBAREA ARCUATA; an ambiguity.
Brassica Napus; no certain record.
Brassica Rapa; always a waif.
Bromus secalinus; localities uncertain.
Bromus commutatus and *racemosus*; little or no difference, and confused with *B. mollis* and *B. secalinus*.

CALAMINTHA NEPETA; confused with the aggregate *C. officinalis*.
Callitriche verna; other species confused with it.
Callitriche platycarpa; localities insufficiently recorded.
Callitriche hamulata; *pedunculata*; misnamed for *C. autumnalis*.
Callitriche autumnalis; confused with the preceding.
Campanula patula; sometimes confused with *C. Rapunculus*.
Campanula Rapunculus; of uncertain occurrence in most of the reported localities.
Campanula latifolia; confounded in the records with *C. Trachelium*.
Carduus Marianus; a garden escape.
Carex bœnninghauseniana; a most unsatisfactory species; some examples approximating to *C. axillaris*, others to depauperised *C. paniculata*.
Carex Œderi; reports of occurrence included with *C. flava*.
Carex lævigata; confused with *C. binervis*.
Ceratophyllum aquaticum (agg.); respective localities of the two species imperfectly separable.
Cicuta virosa; many erroneous records.
Convallaria majalis; introduced in some of its localities.
Crepis biennis and *C. taraxacifolia* have been confused together.

Cuscuta europæa; much confused with the other species.
Cynoglossum montanum; misrecords occur, from *C. officinale* being mistaken for it.

DAPHNE MEZEREUM; a garden escape in many places; difficult to decide where native.
Doronicum Pardalianches; records partly errors.

ELYMUS ARENARIUS; confused with *Ammophila* (*Psamma*).
Epilobium roseum; often overlooked or misnamed.
Epilobium obscurum; more frequent than *E. tetrapterum*.

FESTUCA ELATIOR; confused with *F. pratensis*.
Festuca rubra; confused with the variety *duriuscula* and with *F. ovina*.
Filago apiculata and *spathulata*; distribution imperfectly known.
Fragaria elatior; records partly errors; other garden stragglers reported under this name.
Fumaria capreolata (agg.); no separate record of segregates.

GALIUM ANGLICUM; many doubtful localities on the records.

HABENARIA BIFOLIA and *chlorantha*; localities much confused.
Helianthemum surrejanum; extinct except in gardens.
Helleborus fœtidus; a doubtful native.
Helleborus viridis; frequently a garden escape.
Herminium Monorchis; extinct or mistaken in several counties.
Hesperis matronalis; nowhere permanent; a garden waif.
Hypericum tetrapterum; segregates confused by false naming.

IMPATIENS NOLI-ME-TANGERE; localities misreported, but may be native.

JUNCUS DIFFUSUS; a dubious species, usually, or always, infertile.

KNAPPIA AGROSTIDEA; extinct in Essex.

LAMIUM INCISUM; often confounded with *L. purpureum*.
Lotus tenuis; habitats, apart from *L. corniculatus*, insufficiently reported.
Luzula multiflora; confused in the books with *L. campestris*.

MARRUBIUM VULGARE; often a casual.
Mentha piperita; distribution uncertain.
Myosotis cæspitosa; much confused with *M. palustris*.
Myriophyllum alternifolium, and *M. spicatum*; confused with one another; localities of the former not fully established.

NARCISSUS PSEUDO-NARCISSUS; several recorded localities belong to *N. major* escaped from gardens.
Nephrodium dilatatum; strangely confused with *N. spinulosum*.

ŒNANTHE PIMPINELLOIDES; many false localities on record.

Œnanthe Lachenalii; named *Œ. pimpinelloides* in old books.
Œnanthe silaifolia; many false localities.
Œnanthe fluviatilis; not distinguished from *O. Phellandrium* by the foreign botanists.
Onobrychis sativa; native on the chalk.
Onopordum Acanthium; often a casual or alien.
Ophrys aranifera; apparently decreasing (not seen near London).
Orchis militaris, *Simia*, and *purpurea*; localities of these much confused.
Orchis hircina; very rare.
Orobanche elatior; several localities erroneous.

PEUCEDANUM OFFICINALE; *Œnanthe Lachenalii* mistaken for it occasionally.
Pinus sylvestris; nowhere truly wild but in the Highlands.
Polygonatum multiflorum; difficult to decide where truly native.
Potamogeton heterophyllus; confused with *P. obtusifolius*, *natans*, and *polygonifolius*.
Potamogeton natans; often confused with the preceding.
Potamogeton polygonifolius; often mislabelled *P. natans*.
Polygonum mite; allied to *P. Persicaria*, but has been confused with *P. minus*.
Polystichum aculeatum and var. *lobatum*; confused with *P. angulare*.
Populus alba; impossible to fix its native area.

ROSA RUBIGINOSA; true and false localities inextricably confused.
Rubus (*fruticosus*); distribution that of a name only, as regards its segregate forms.

SALVIA PRATENSIS; many erroneous localities (extinct at Cobham?).
Scirpus sylvaticus; occasionally mistaken for *S. carinatus*.
Scirpus multicaulis; partly confused with *S. uniglumis*.
Scirpus cæspitosus; occasionally mislabelled *S. pauciflorus*.
Scirpus carinatus; an unsatisfactory species; localities confused with those of *S. triqueter* and *S. glaucus*.
Sclerochloa Borreri; an unsatisfactory species.
Scrophularia Ehrhardti, and *Balbisii*; confused under the common name of *S. aquatica*.
Senecio viscosus; localities decreasing.
Silene quinquevulnera; a casual.
Sisymbrium Irio; localities uncertain.
Sonchus palustris; erroneously reported through misnamers for *S. arvensis*.
Sonchus oleraceus; *S. asper* often confused with this.
Stachys ambigua; name often misapplied to *S. palustris*.
Statice bahusiensis; doubtfully, sufficiently, or permanently distinct from *S. Limonium*.

TARAXACUM PALUSTRE; not always distinguished from *T. officinale*.
Trifolium ochroleucum; many localities need verification.

ULEX NANUS; distribution ill-ascertained.
Utricularia intermedia; several false localities.
Utricularia minor; often mislabelled *U. intermedia.*

VALERIANELLA AURICULA; perhaps often overlooked.
Verbascum pulverulentum; *V. nigrum* and *V. Lychnitis* have been mistaken for it in some counties.
Verbascum Blattaria; often temporary.
Verbascum virgatum; mostly a casual.
Vinca minor; in Britain usually a stray from gardens.
Viola calcarea; a dwarf state of *V. hirta.*
Viola permixta; inconveniently intermediate between *V. hirta* and *V. odorata.*
Viola sylvatica; the var. *Riviana,* the usual representative of *V. canina;* old records refer to this only.

LOCALITIES.

1. HAMPSTEAD HEATH.

A road from Hampstead to Highgate, which leads along the crest of the northern heights, separates what is called East Heath from West Heath. Height of the ridge, over 400 feet; subsoil, sand. East Heath is a grassy common, with a pond above, and some reservoirs at the lower part of it. West Heath slopes to the north-west, and is divided by a road through North End to Hendon, into two portions. The sand which caps the formation is most apparent in the section nearest Highgate, and here the ground is much broken into pits and hillocks. In the hollow of the further slope is a bog, which drains into a pool below. Much Bracken here, with a few white- and black-thorns.

*Achillæa Ptarmica.
*Aira præcox; *caryophyllea; *cæspitosa; *flexuosa.
*Alisma Plantago (ponds).
*Anemone nemorosa (among the Fern).
*Anthemis nobilis.
*Arenaria rubra.
*Ballota nigra (Frognal road, etc.).
*Bartsia Odontites (lane leading to Fortune Green).
*Bunium flexuosum (lower part).
*Calluna vulgaris.
*Caltha palustris.
*Campanula rotundifolia.
*Carduus palustris.
*Carex flava; *ovalis; *panicea; *pilulifera; *stellulata; *vulgaris.
*Draba vernalis (walls about).
*Drosera rotundifolia (bog).
*Elodia canadensis (pond).
*Epilobium montanum.
*Erica cinerea; *Tetralix.
*Eriophorum angustifolium.
*Euonymus europæus (lane leading to Fortune Green. One tree).
*Festuca ovina.
*Galium saxatile; *uliginosum.
*Genista anglica.
*Glyceria fluitans (in water-holes).
*Hieracium Pilosella; *vulgatum; *umbellatum.
*Hydrocotyle vulgaris.
*Hypericum pulchrum.
*Hypochœris radicata.
*Jasione montana.
*Juncus obtusiflorus; *bufonius; *uliginosus; *acutiflorus (in a pond, right of lane leading to Fortune Green; a tuft or two).
*Lactuca muralis (wall by the 'Spaniards').
*Leontodon hirtus; *autumnalis.
*Linaria Cymbalaria (walls near).
*Lotus major; *corniculatus.
*Luzula campestris; pilosa.
*Lychnis Flos-cuculi (in the bog).
*Menyanthes trifoliata (bog).
*Molinia cærulea.
*Montia fontana.
*Nardus stricta.
*Nasturtium officinale; *terrestre (banks of pool, right of the lane leading to Fortune Green).
*Ononis spinosa (East Heath, near the lower reservoirs, and North End).
*Ornithopus perpusillus.
*Oxalis Acetosella.
*Pedicularis sylvestris; *palustris.
*Polygonum Hydropiper (in holes).
*Potentilla Fragariastrum; *reptans; *Tormentilla.
*Pyrus Malus.
*Ranunculus Flammula; *hederaceus.
*Rumex Acetosella.
*Sagina procumbens; *apetala (walls about).
*Salix repens; *fragilis (the latter are probably all planted trees).
*Sarothamnus scoparius.
*Saxifraga tridactylites (walls).

*Scabiosa succisa.
*Scilla nutans (among the Fern).
*Scirpus palustris; *setaceus.
*Scutellaria minor (about the pond).
*Senebiera Coronopus.
*Senecio Jacobæa.
*Serratula tinctoria (among the Fern).
*Solanum nigrum (lanes about).
*Solidago Virgaurea (among the Fern).
*Stachys sylvestris (ditch below).
*Stellaria uliginosa (bog).
*Teucrium Scorodonia.
*Triodia decumbens.
*Ulex europæus.
*Vaccinium Myrtillus (near the Fir-trees by the 'Spaniards').
*Vicia Cracca (near the 'Spaniards').
*Viola sylvestris; *palustris.

CRYPTOGAMS.

*Asplenium Ruta-muraria (wall near Highgate).
*Aulocomnion palustre (bog).
*Bartramia fontana (a tuft or two, upper part of the bog).
*Bryum cæspitium; *atro-purpureum.
Campylopus torfaceus (scarce).
*Ceratodon purpureus.
*Chara vulgaris (pond right of lane leading to Fortune Green).
*Dicranella cerviculata; heteromalla.
*Dicranum scoparium.
*Equisetum arvense; *limosum; *palustre.
*Fissidens bryoides (sides of ditch below).
*Hypnum cuspidatum; *cordifolium; *confertum; *cupressiforme (also a white-leaved variety of this); *fluitans; *denticulatum; *populeum;* rutabulum; *stramineum (a tuft or two in the bog);* murale; *velutinum.
*Jungermannia connivens; *crenulata; *turbinata; scalaris; *Trichomanes; *multifida; *bicuspidata; *heterophylla.
*Mnium hornum (sides of ditch below).
*Pleuridium subulatum.
*Polytrichum commune; *piliferum.
*Pottia truncata (lane leading to Fortune Green).
*Pteris aquilina (abundant).
*Sphagnum cymbifolium; acutifolium.

APPENDIX.

*Bunium flexuosum; *Melampyrum pratense (Highgate wood).
Herniaria hirsuta ? (fields near Colney Hatch, *olim*).
*Lathyrus Nissolia (roadside Finchley road, a mile or so from North End).
*Mercurialis annua (gardens &c., at Child's Hill).
Polygonum Bistorta (borders of Bishop's wood, below; and in the meadow, now rare).
Populus tremula; Chrysosplenium oppositifolium (Bishop's wood, enclosed).
*Rhinanthus Crista-Galli (meadows near North End).
Senebiera didyma (between Southgate and Colney Hatch, by the roadside).
Smilacina bifolia; Caen wood (enclosed)
Stachys arvensis (cornfield in the wood, formerly).

2. BARNES COMMON.

The Thames, in its course of three or four miles from Mortlake to Putney, makes a considerable curve, and the loop of land thus enclosed, constitutes, in a great degree, the suburban district of Barnes. The common is situated at the base of the peninsula, and is an open level space more or less covered with Furze, Broom, Briars, Bracken, and Heath; with a gravelly subsoil, and a pool or two of water. It is crossed by roads, and by a branch of the South-Western Railway. On its northern limits is a ditch, and a narrow strip of marshy pasturage. There are two mortuary chapels upon it, with their respective cemeteries; besides two or three recently erected buildings with their enclosures. Distance: one mile from Hammersmith Bridge; by the roadside, Nasturtium sylvestre, abundant; adjoining cultivated land, Fumaria officinalis, &c.

ON THE COMMON.

*Acorus Calamus.
*Aira flexuosa; *præcox.
*Alchemilla arvensis.
*Anthemis nobilis.
*Arenaria rubra.
*Campanula rotundifolia.
*Carex pilulifera; *flava; *hirta; *vulgaris; *Pseudo-cyperus (ditch).

*Centaurea Calcitrapa (towards Putney).
*Dipsacus sylvestris.
*Epilobium hirsutum.
*Erodium cicutarium.
*Erysimum cheiranthoides (near Mortlake).
*Galium verum; *saxatile; *palustre; *uliginosum.
*Genista anglica.
*Glyceria aquatica.
*Helosciadium nodiflorum; *inundatum.
*Hieracium vulgatum; *Pilosella.
*Holcus mollis; *lanatus.
*Hydrocharis Morsus-ranæ.
*Jasione montana.
*Juncus lamprocarpus, &c.
*Leontodon hirtus; *hispidus; *autumnalis.
*Lotus corniculatus; *major.
*Lythrum Salicaria.
*Medicago maculata.
*Mœnchia erecta.
*Montia fontana.
*Nasturtium sylvestre; *amphibium.
*Œnanthe fistulosa.
*Ononis spinosa.
*Ornithopus perpusillus.
*Pedicularis sylvestris.
*Polygala vulgaris.
*Potentilla Tormentilla.
*Ranunculus Flammula.
*Rosa spinosissima (a few shrubs near some newly-built houses[1]).
*Rumex Hydrolapatha.
*Salix repens.
*Scleranthus annuus (a few plants).
*Senecio sylvestris.
*Spiræa Ulmaria.
*Teesdalia nudicaulis (left of road to Roehampton).
*Thymus Serpyllum.
*Trifolium subterraneum; filiforme.

MEADOWS NEAR THE COMMON.

*Populus nigra, ♂

CRYPTOGAMS.

*Aulocomnion palustris.
*Bartramia fontana (a patch or two).
*Bryum cæspitosum.
Campylopus torfaceus, &c.

3. BANKS OF THE THAMES FROM PUTNEY TO KEW.

A towing and footpath leads along the banks of the river on the Surrey side. The Thames hereabouts is within full tidal influence, and the rise and fall is considerable; however, the brackish water has no injurious effect upon the fresh-water vegetation. On the Middlesex side the lands are all enclosed.

BY THE RIVER-SIDE AND BORDERING DITCHES.

*Angelica sylvestris.
*Barbarea vulgaris.
*Bromus racemosus.
*Caltha palustris.
*Carex riparia; *paludosa; *acuta (this in a marshy enclosure near Putney).
*Chærophyllum Anthriscus; *sylvestre.
*Digraphis arundinacea.
*Epilobium hirsutum.
*Geranium pratense (Osier holts beyond Mortlake[2]).
*Geranium pyrenaicum (railway bank near Mortlake).
*Glyceria aquatica.
*Helosciadium nodiflorum.
*Humulus Lupulus (incidental).
*Iris Pseudacorus.
*Lepidium Draba (a patch of this near the factory[3]).
*Lythrum Salicaria.
*Melilotus officinalis.
*Myosotis palustris.
*Nasturtium officinale; *sylvestre; *amphibium; *terrestre.
*Œnanthe crocata (plentiful).
*Papaver dubium (walls, Mortlake).
*Parietaria diffusa (walls, Mortlake).
*Populus nigra, ♀
*Petasites vulgaris.
*Ranunculus Ficaria.
*Salix fragilis; *alba; *triandra; *viminalis; *cinerea.
*Sambucus nigra.
*Scrophularia Balbisii (aquatica).
*Sium latifolium (one old root).
*Spiræa Ulmaria.
*Symphytum officinale.

[1] On remains of what was once a hedgebank, where it was probably planted.
[2] These holts have been converted into market gardens, but the bordering ditches may be still productive.
[3] ? From seed washed up from the coast of Thanet, where the plant is extremely common.

*Tanacetum vulgare.
*Valeriana officinalis.
*Veronica Anagallis.

*Leskea polycarpa (old posts and stumps near Kew).

ROADSIDE BETWEEN BARNES AND KEW.

*Ægopodium Podagraria (about Mortlake).
*Anthriscus vulgaris.
*Lychnis vespertina.
*Medicago maculata (banks of the reservoir).
*Potentilla argentea (banks of the reservoir, scarce).
*Sedum acre (banks of the reservoir).
*Silene inflata (banks of the reservoir).
*Trifolium filiforme (banks of the reservoir).
*Tussilago Farfara (banks of the reservoir).
*Valerianella olitoria (banks of the reservoir).
*Viola odorata (hedges towards Kew).
*Scirpus lacustris; *maritimus; *palustris: (cultivated lands between Mortlake and Sheen), *Æthusa Cynapium; *Papaver Rhœas; *Fumaria officinalis; muralis (capr. agg.).

4. PUTNEY HEATH AND WIMBLEDON COMMON

Are continuous with each other, and form a somewhat elevated and extensive plateau. The heath is crossed by the road from Wandsworth to Kingston. The common has been converted into a rifle-practising ground; but the slope westwards is covered with scrub of stunted Oak, Hazel, Birch, and Sallows, intertangled with tall Furze, and broken by three or four ravines; the valley below is traversed by a sluggish brook. The most considerable of these ravines fronts a windmill. A drain has been recently cut through it, to the prejudice of its flora; to some extent, at least. At the further extremity of the plateau is an ancient entrenchment; near this, Coombe Wood, enclosed. On Putney Heath are several buildings of recent construction, but much scrub is still standing; Furze, Briar, Honeysuckle, Heath, and Bracken; with scattered Birch, Elm, and White-thorn. Near Roehampton are some pools, and tracts of marshy ground, sandpits, &c.; subsoil, gravel.

*Achillea Ptarmica.
*Alisma Plantago; *ranunculoides.
*Anthemis nobilis.
*Bunium flexuosum.
*Caltha palustris.
*Carduus pratensis.
*Carex panicea; *glauca; *hirta; *vulgaris; *disticha (ravine); *teretiuscula (ravine); *ovalis; *stellulata; *binervis; *flava; *remota; *stricta (one patch, Putney Heath); *pilulifera.
Claytonia perfoliata (turf bank near the mill).
*Cuscuta Epithymum.
*Drosera rotundifolia.
*Epilobium montanum; *tetragonum; *hirsutum; palustre.
*Erica cinerea; *Tetralix.
*Eriophoron angustifolium.
*Euonymus europæus.
*Euphrasia officinalis.
*Filago minima (gravelly places).
*Galium Mollugo; *palustre; *uliginosum.
*Genista anglica.
*Hieracium vulgatum; *Pilosella; *umbellatum.
*Hypericum perforatum; *pulchrum; humifusum; *quadrangulare.
*Inula dysenterica.
*Juncus obtusifolius; *lamprocarpus; *squarrosus; *supinus; *bufonius.
*Lysimachia nemorum.
*Lythrum Salicaria.
*Medicago maculata (gravel pits).
*Menyanthes trifoliata (ravine).
*Molinia cærulea.
*Myosotis arvensis; *palustris; *collina; *versicolor.
*Narthecium ossifragum (near Roehampton).
*Œnanthe fistulosa.
*Orchis latifolia (further ravine).
*Peplis Portula.
*Potamogeton natans; polygonifolius.
*Rhamnus Frangula.
*Rosa stylosa.
Rubus suberectus; *rhamnifolius; *carpinifolius; rudis; rosaceus; *glandulosus.

*Sagittaria sagittifolia (by the stream).
*Salix repens; *Caprea; *aurita; *cinerea (by the stream).
*Scirpus fluitans; *setaceus.
*Scutellaria minor.
*Serratula tinctoria.
*Sonchus arvensis (by the stream).
*Sparganium simplex (by the stream).
*Spiræa Ulmaria (by the stream).
*Stellaria uliginosa.
*Triodea decumbens.
*Valeriana dioica (ravine).
*Veronica scutellata.

CRYPTOGAMS.

*Bartramia fontana (ravine).
*Dicranella cerviculata.
*Hypnum stellatum (a patch or two); *cordifolium (scarce).
*Marchantia polymorpha.
*Mnium punctatum (ravine).
*Polytrichum commune.
*Sphagnum obtusifolium.
*Webera albicans; *nutans.
*Weissia cirrhata (old railings, Roehampton Lane).

APPENDIX.

*Carduus palustris (Coombe Wood).
*Carex sylvatica (Coombe Wood).
*Digitalis purpurea (Coombe Wood).
*Nephrodium Filix-mas (Coombe Wood).
*Primula vulgaris (Coombe Wood).
*Scilla nutans; also with white variety (Coombe Wood).
*Brachypodium sylvaticum (borders).
*Daucus Carota (borders).
*Chenopodium olidum (Putney bottom by the roadside, and in waste places).
*Erysimum cheiranthoides (Putney bottom by the roadside, and in waste places).
*Euphorbia Helioscopia (Putney bottom by the roadside, and in waste places).
*Thlaspi arvense (Putney bottom by the roadside, and in waste places).
*Verbena officinalis (Putney bottom by the roadside, and in waste places).
*Carex præcox (Richmond Park).
*Corydalis claviculata (hedgebank beyond old camp).
*Rhinanthus Crista-Galli (meadow below the mill).
Anemone apennina (Wimbledon Park).
Ornithogalum umbellatum (Wimbledon Park).
Symphytum tuberosum (Wimbledon Park).
Tulipa sylvestris ? (Wimbledon Park).

5. LANES AND ROADSIDES ABOUT HENDON, NEASDON, AND KINGSBURY.

These suburban localities are situated in the hollow, drained by the Brent; an offshoot, as it were, of the great basin of the Thames. The subsoil, clay; or at a greater or less depth, gravel drift, resting upon clay. The lands are mostly meadow. To the northwards are the heights of Stanmore and Harrow Weald. South is the Hampstead and Highgate ridge. Near Neasdon a dam has been built across the Brent, and a tract of low-lying land above it converted into an extensive lake. In the shady lanes hereabouts, and by the banks of this sluggish stream, the under-mentioned plants may be procured.

*Æthusa Cynapium.
*Aira cæspitosa.
*Agrimonia Eupatoria.
*Ajuga reptans.
*Allium ursinum.
*Arum maculatum.
*Avena flavescens; *elatior.
*Ballota nigra.
*Bartsia Odontites.
*Bidens cernua; tripartita.
*Bromus asper; *racemosus.
*Carex remota; *hirta; *vulpina; *divulsa; *panicea.
*Chærophyllum Anthriscus; temulentum.
*Conium maculatum (hollow below the railway-bridge between Willesden and Neasdon; old disused lane between Kingsbury and Stanmore Marsh).
*Cynosurus cristata (Hendon, plenty).
*Epilobium montanum.
*Fraxinus excelsior (roadsides).
*Galeopsis Tetrahit.
*Galium Aparine.
*Geranium molle; pusillum; Robertianum.
*Heracleum Sphondylium.
*Hottonia palustris (in some pools near the reservoir, near Woodford House).
*Hypericum perforatum.
*Inula dysenterica.

*Juncus glaucus, &c.
*Lactuca virosa (plentiful about the railway-bridge near Neasdon; also in lane leading from Neasdon to the Edgware Road).
*Lapsana communis.
*Lathyrus pratensis.
*Linaria vulgaris.
*Lychnis Githago; *diurna.
*Medicago lupulina (railway banks).
*Melica uniflora (near Hendon).
*Mentha hirsuta; *sativa.
*Mercurialis perennis.
*Milium effusum (Brondesbury).
*Nasturtium sylvestre (Edgware Road by the reservoir, plentiful).
*Nuphar lutea (in the Brent).
*Œnanthe silaifolia (in the ponds north of the reservoir; near Woodford House).
*Pimpinella Saxifraga.
*Polygonum Hydropiper; amphibium.
*Pyrus Malus.
*Ranunculus bulbosus (meadows); *auricomus (lanes, Kingsbury).
*Rhamnus catharticus (by the Brent).
*Rosa arvensis; systyla; canina.
*Salix Caprea; *cinerea; *fragilis; *viminalis.
*Scilla nutans (near Whitchurch especially, and Woodford House).
*Scirpus palustris; sylvaticus (Brent).
*Scrophularia Balbisii; nodosa.
*Silaus pratensis.
*Sison Amomum (Brondesbury, plenty).
*Solanum nigrum.
*Stachys sylvestris.
*Stellaria Holostea; graminea.
*Torilis Anthriscus.
*Trifolium fragiferum (Edgware roadside).
*Triticum caninum, with other common grasses.
*Ulmus suberosus (roadsides).
*Veronica Chamædrys; *officinalis; hederifolia.

CRYPTOGAMS.

*Anomodon viticulosus (stump of a tree, Kingsbury Lane).
*Dicranum bryoides (Hole Lane).
*Equisetum limosum (ponds near the reservoir)
*Fissidens taxifolius (Hole Lane).
*Homalia trichomanes (Hole Lane).
*Hypnum filicioum (Hole Lane).
*Hypnum populeum; *velutinum; *cupressiforme (Kingsbury Lane).
*Hypnum serpens, &c.; *speciosum [var. of prælongum] (stone wall of a small drain near Neasdon).
*Jungermannia complanata; *bidentata, &c. (Hole Lane).
*Mnium undulatum (Hole Lane, plenty, but not fruiting).
*Pottia truncata (hedgebanks), &c., &c.
*Tortula ruralis (thatch at Cricklewood).
*Weissia cirrhata (thatch at Cricklewood).

6. WILLESDEN: BANKS OF THE BRENT, AND PADDINGTON CANAL.

Situated in the basin of the Thames, with a deep gravelly subsoil, resting upon clay; water drainage to the Brent stream; pasturage, the prevailing agricultural characteristic; a corn or fodder-field, occasionally only; country level or nearly so.

*Achillea Ptarmica; Millefolium.
*Agrimonia Eupatoria.
*Alisma Plantago.
*Ballota nigra.
*Bartsia Odontites.
*Bidens cernua.
*Brachypodium sylvaticum (lane).
*Butomus umbellatus.
*Carduus arvensis; *crispus; lanceolatus.
*Carex vulpina; *riparia; *teretiuscula (one hassock of this in canal above Willesden).
*Centaurea Jacea; *Cyanus (in cornfield, scarce).
*Chærophyllum Anthriscus; *sylvestre.
*Chenopodium olidum (by the railway bridge, Harrow Road).
*Daucus Carota.
*Erodium cicutarium.
*Euonymus europæus (hedge, Acton Road).
*Galeopsis Tetrahit.
*Geranium pusillum; *Robertianum.
*Geum urbanum.
*Helminthia echioides.
*Helosciadium nodosum.
*Hieracium Pilosella.
*Hydrocharis Morsus-ranæ (canal).
*Hypochœris radicata.
*Inula dysenterica.
*Juncus communis; *glaucus.
*Lactuca virosa (plentiful by the railway-bridge, Acton road).
*Leontodon hispidus; *autumnalis.
*Linaria vulgaris.
*Lychnis Githago.

*Lycopus europæus.
*Lysimachia vulgaris; Nummularia (by the Brent, scarce).
*Mentha hirsuta.
*Myosotis arvensis.
*Nymphæa alba; lutea (in the Brent).
*Ononis spinosa.
*Phragmites communis (ditch by a cornfield).
*Pimpinella Saxifraga.
*Potamogeton crispus; *pusillus; *pectinatus.
*Prunella vulgaris.
*Reseda luteola.
*Rosa arvensis; canina.
*Sagittaria sagittifolia (in the Brent).
*Scabiosa succisa.
*Scirpus sylvaticus.
*Scutellaria galericulata.
*Silaus pratensis.
*Sinapis alba; arvensis.
*Sison Amomum (railway-bridge, Acton road).
*Solanum nigrum; *Dulcamara.
*Sonchus arvensis (cornfield, Willesden).
*Sparganium ramosum; *simplex.
*Spiræa Ulmaria.
*Stachys palustris; *sylvestris.
*Stellaria Holostea; *graminea.
*Thlaspi arvense (cornfield, Willesden).
*Trifolium procumbens.
*Tussilago Farfara.
*Veronica Buxbaumii (cornfield, Willesden).
*Vicia sepium; *hirsuta; *tetrasperma (waste ground, where the Brent crosses the canal).

CRYPTOGAMS.

*Leskea polycarpa (trunks of trees by the Brent, at Stonebridge).
*Orthotrichum affine (trunks of trees by the Brent at Stonebridge).

APPENDIX.

The Brent, about Greenford and Perivale.

*Acorus Calamus, &c.
Iris fœtidissima ? (pastures about Perivale).
*Lysimachia vulgaris (scarce).
*Nuphar lutea.
*Nymphæa alba (scarce).
*Ranunculus aquatilis (agg.).
*Sagittaria sagittifolia.
*Thalictrum flavum.

ABOUT BRONDESBURY.

*Atriplex deltoidea; *angustifolia.
*Avena elatior (hedges).
*Bromus sterilis.
*Chenopodium album; rubrum.
*Erysimum cheiranthoides.
*Hordeum murinum.
*Plantago lanceolata.
*Rumex obtusifolius; *conglomeratus; *crispus.
*Senebiera Coronopus.

7. PASTURES AND LANES ABOUT TOTTENHAM AND EDMONTON; BANKS OF THE LEA, AND OF THE LEA CANAL.

Low-lying and damp, but well-drained meadows border the river Lea, and the canal which runs by the side of it, more or less in close proximity. Low hills at no great distance extend upwards on either hand, in a northerly direction. The subsoil is either gravel or alluvium.

*Acer campestre (hedges).
*Ægopodium Podagraria.
*Æthusa Cynapium.
*Agrimonia Eupatoria.
*Alisma Plantago.
*Alnus glutinosa (Chingford Mills, &c.).
*Alopecurus pratensis; *agrestis (cornfields, Edmonton); *geniculatus; *fulvus (gravel pit, Tottenham).
*Angelica sylvestris (ditches).
*Anthoxanthum odoratum.
*Arctium minus.
*Artemisia vulgaris (canal banks).
*Bidens cernua; *tripartita.
*Bryonia dioica (hedges).
*Butomus umbellatus (ditches).
*Caltha palustris.
*Cardamine pratensis.
*Carduus arvensis; *lanceolatus.
*Carex riparia; paludosa.
*Ceratophyllum aquaticum.
*Chærophyllum Anthriscus; *sylvestre; *temulum (hedges).
*Chrysanthemum Leucanthemum.
*Circæa lutetiana (White Hart Lane).
*Conium maculatum (hedge in a field near Park station).
*Dactylis glomerata.
*Digraphis arundinacea.
*Dipsacus sylvestris (canal banks).
*Elodia canadensis (canal).
*Epilobium hirsutum; *parviflorum; *montanum; *tetragonum.
*Festuca pratensis.
*Fumaria officinalis (cultivated fields, about).

*Galium palustre (ditches).
*Geranium dissectum; *molle; *pusillum; *Robertianum.
*Geum urbanum (lanes).
*Glyceria aquatica.
*Hedera Helix (Chingford Old Church).
*Helosciadium nodosum (ditches).
*Heracleum Sphondylium.
*Hypochœris radicata.
*Inula dysenterica.
*Iris Pseudacorus.
*Juncus effusus; conglomeratus; *glaucus.
*Lemna trisulca (near Ponder's End, in a field).
*Leontodon hispidus; *autumnalis.
*Lychnis Flos-cuculi; *diurna.
*Lycopus europæus.
*Lythrum Salicaria.
*Malva sylvestris; *rotundifolia.
*Matricaria inodora; *Chamomilla (market gardens).
*Mentha hirsuta; *sativa.
*Myosotis palustris.
*Myriophyllum spicatum (canal).
*Nasturtium officinale.
*Nuphar lutea.
*Œnanthe fistulosa; *Phellandrium; *fluviatile.
*Pastinaca sativa (canal banks).
*Petasites vulgaris (banks of Chingford Mill stream).
*Phleum pratense.
*Pimpinella Saxifraga.
*Poa pratensis; trivialis.
*Potamogeton natans; lucens; *crispus; *pusillus; *pectinatus; *perfoliatus.
*Ranunculus peltatus; *circinatus; *Ficaria, &c.
*Reseda luteola (canal bank).
*Rhinanthus Crista-Galli (meadows).
*Rumex Hydrolapathum; Acetosa.
*Sagittaria sagittifolia.
*Salix alba; *cinerea; *fragilis; *triandra; *viminalis; *Caprea (lanes); rubra? stipularis? Smithiana (Lea Bridge Road, *olim*).
*Scirpus lacustris.
*Scrophularia Balbisii; *nodosa.
*Scutellaria galericulata.
*Senebiera Coronopus.
*Silaus pratensis.
*Sium angustifolium (ditches).
*Solanum Dulcamara (hedges).
*Sparganium ramosum.
*Spiræa Ulmaria.
*Stachys palustris; *sylvestris.
*Stellaria Holostea; *graminea.
*Symphytum officinale.
*Tanacetum vulgare (canal bank).
*Thalictrum flavum (ditches in the flats scarce).
*Torilis Anthriscus.
*Tragopogon pratensis.
*Trifolium procumbens; *minus.
*Tussilago Farfara.
*Typha latifolia; *angustifolia.
*Valeriana officinalis.
*Veronica Anagallis; *Beccabunga.

CRYPTOGAMS.

*Bryum cæspitium; *capillare (walls) *argenteum (banks of canal near lock walls, White Hart Lane); *erythrocarpum (canal bank).
*Ceratodon purpureus.
*Dicranum bryoides (Chingford Mill).
*Didymodon rubellus (walls).
*Equisetum limosum; *arvense.
*Fontinalis antipyretica (canal and mill stream.)
*Funaria hygrometrica.
*Homalia Trichomanes (stumps of trees lane near gasworks).
*Hypnum rutabulum; *velutinum; *confertum; *ruscifolium (mill-stream) *murale; *riparium and *serpens (by the canal); *denticulatum (hedge banks); *populeum; *cuspidatum; *fluitans (marshes near Clapton, fruiting freely); *tenellum (churchyard); *cupressiforme; *prælongum.
*Marchantia polymorpha (wall, Edmonton).
*Nidularia, species (on an old plank by canal).
*Pottia cavifolia.
*Tortula muralis; *unguiculata (canal bank near Tottenham lock).
*Weissia controversa (Chingford road).

8. EPPING FOREST, AND COPSES, LANES, &c., ABOUT CHINGFORD WOODFORD, AND WALTHAMSTOW.

Much clearing, enclosing, and building, has taken place in these neighbourhoods within the last forty years; and many plants recorded by Forster, as growing freely there, are no longer to be found. They have become, and justly so, favourite suburban places of residence. Viewed in its entirety, the forest comprises the extensive wooded upland running in a north-easterly direction which separates the vale of the lower Lea and affluent from that of the Roding, into which the slopes drain on either side. The subsoil is gravel. It is intersected by the high road, from Lea Bridge, through Woodford to Epping and Newmarket, into which run cross-roads from Walthamstow and Chingford. Woodford occupies a sort of clearing in the forest, which may be described as hereabouts consisting of detached patches, partly enclosed. These, together, with the outlying copses known as Lark's Wood and Hawkswood, produce a few woodland plants, but the localities are so much frequented, that little out of the common may be expected. Oak and Hornbeam appear to be most prevalent in the drier parts. In the hollow below Woodford, which is of a somewhat marshy character, is a dense scrub of White-thorn, Black-thorn, Briar, Bramble, Holly, Birch, Sallow, dwarf Oak, Hazel, and Bracken.

*Allium ursinum (lane, Chingford Hatch).
*Anemone nemorosa.
*Bunium flexuosum.
*Calluna vulgaris.
*Carex glauca; *panicea; *flava.
*Cornus sanguinea.
*Euphorbia amygdaloides.
*Euphrasia officinalis.
*Fragaria vesca.
*Galium Mollugo (hedges).
*Genista anglica.
*Helminthia echioides.
*Hieracium Pilosella.
*Hottonia palustris (pools by the road-side, near Woodford).
*Hydrocotyle vulgaris.
*Lamium Galeobdolon (scarce).
*Lonicera Periclymenum.
*Lotus corniculatus.
*Lysimachia Nummularia.
*Neottia Nidus-avis.
*Orchis maculata (marshy hollow).
*Ornithopus perpusillus.
*Orobanche major (scarce).
*Orobus tuberosus.
*Oxalis Acetosella.
*Pedicularis sylvestris.
*Potentilla reptans; Fragariastrum; *Tormentilla; *anserina.
*Primula vulgaris.
*Tamus europæus.
*Thymus communis (molehills).
*Salix repens; *Caprea; *aurita.
*Sanicula europæa.
*Scilla nutans.
*Stellaria Holostea; graminea.
*Veronica officinalis; *Chamædrys.
*Viburnum Opulus (lanes).
*Viola odorata; *sylvestris.

CRYPTOGAMS.

*Atrichum undulatum.
*Aulocomnion palustre; *androgynum.
*Ceratodon purpureus.
*Dicranum undulatum.
*Didymodon rubellus (walls, Walthamstow).
*Funaria hygrometrica.
*Grimmia pulvinata (walls, Walthamstow).
*Homalia trichomanoides (Chingford Hatch.
*Hypnum purum; *piliferum (roadside beyond Woodford); *cupressiforme; *cordifolium (between Woodford and Walthamstow); *cuspidatum; *denticulatum (hedgebanks); *populeum; *prælongum; *rutabulum; *velutinum; *Schreberi; *serpens; *glareosum; *splendens; *squarrosum; *triquetrum; *loreum; *fluitans; *aduncum; *tenellum; *striatum.
*Jungermannia complanata (trunks of trees); *bidentata (banks).
*Marchantia hemisphærica (sides of a pit).
*Mnium hornum (near Woodford).
*Mnium undulatum (Chingford Hatch).
*Nephrodium Filix-mas.
*Orthotrichum diaphanum (Chingford).
*Pleuridium subulatum.
*Polytrichum commune; *piliferum; *juniperum.
*Pottia truncata.
*Sphagnum cymbifolium; *acutifolium.
*Thuidium tamariscinum.
*Webera nutans (gravel pits).
*Weissia controversa.

APPENDIX.

Larkswood; Hawkswood.

Calamagrostis Epigejos.

9. BLACKHEATH, AND THE MARSHES BELOW GREENWICH.

Many plants grew formerly upon Blackheath, and in the marshes between Greenwich and Woolwich. The heath is now nothing but a grassy common, trodden by thousands of persons, and enclosed on all sides, but the park, by suburban villas of recent construction. The subsoil is gravel, and the surface in many places is pitted with depressions, whence this material has been dug out. These hollows may be examined, especially a large pit at the north-east corner of the plain, whence the road leads down to the flats by the Thames; elevation, 140 feet.

The marshes have long since been drained, enclosed, and converted into market gardens. The intersecting ditches by the roadside, and riverside, however, still afford a few plants characteristic of the locality; subsoil, alluvial.

ON THE HEATH.

*Aira flexuosa; *præcox; *caryophyllæa.
*Alchemilla arvensis.
*Anthoxanthum odoratum.
*Arenaria serpyllifolia.
*Avena flavescens.
*Bellis perennis.
*Campanula rotundifolia.
Carduus acaulis.
Cerastium tetrandrum; *semidecandrum.
Draba verna.
*Erodium cicutarium.
Festuca sciuroides.
*Filago minima.
*Geranium molle.
*Gnaphalium uliginosum.
*Hieracium Pilosella.
*Jasione montana.
Kœleria cristata.
*Leontodon hispidus.
*Luzula campestris.
*Mœnchia erecta.
Myosotis versicolor.
*Ornithopus perpusillus.
*Polygala vulgaris.
*Potentilla Tormentilla; anserina; reptans.
*Rumex Acetosella.
*Sagina apetala; *procumbens.
Scilla autumnalis.
Sedum acre.
*Senebiera Coronopus.
*Spergularia rubra.
*Thymus Serpyllum.
Trifolium subterraneum; *filiforme; scabrum.
*Veronica arvensis; *agrestis; *serpyllifolia.

MARSHES AND RIVER-SIDE.

*Apium graveolens.
*Aster Tripolium.
*Catabrosa aquatica.
*Cochlearia anglica.
*Digraphis arundinacea.
*Festuca elatior.
*Glaux maritima.
*Glyceria aquatica.
*Phragmites communis.
*Scirpus maritimus.
*Spergularia neglecta.

CRYPTOGAMS.

Ulva (two or three species).

APPENDIX.

*Corydalis lutea (wall by the road leading from N.E. corner of heath to the flats below).

BY THE RAVENSBOURNE NEAR LEWISHAM.

Cardamine amara; Hesperis matronalis; olim.

10. CHARLTON WOOD, AND CHALK-PIT, WOOLWICH SAND-PITS; SHOOTER'S HILL.

Not much is left standing or unenclosed of what were once known as Charlton and Hanging Woods. Some old chalk-pits near the station, and some sand-pits not far from Woolwich, together with a bordering copse or two, are all that remain; formerly this locality was a very productive one, and the list of plants given by Mr. Cooper in his 'Flora' is a very full one; though many are upon ancient authority—Pamplin, Blackstone, 'Botanist's Guide,' Milne, and others; since their time great changes have occurred here, as elsewhere in the neighbourhood of London. The localities are on the flank of the range of hills which border the flats by the Thames on the

Kentish side, extending from Blackheath to Erith, and attaining an elevation at Shooter's Hill of over 400 feet; subsoil, sand, gravel, overlying chalk.

Anchusa arvensis.
Artemisia vulgaris.
Bryonia dioica.
Calamintha Clinopodium.
Carex depauperata (?) olim.
Centaurea Scabiosa; nigra.
Circæa lutetiana.
Clematis Vitalba.
Cornus sanguinea.
Daucus Carota.
Diplotaxis tenuifolia.
Echium vulgare.
Erigeron acris.
Erythræa Centaurium.
Euphorbia amygdaloides.
Filago germanica.
Galium verum; Mollugo.
Geranium dissectum.
Geum urbanum.
Heracleum Sphondylium.
Hypericum perforatum; hirsutum.
Inula Conyza.
Lactuca Scariola (?) (sandpits).
Lathyrus pratensis.
Lepidium campestre.
Ligustrum vulgare.
Linaria vulgaris.
Lotus corniculatus.
Medicago maculata; lupulina.
Mercurialis perennis.
Ononis arvensis.
Origanum vulgare.
Pastinaca sativa.
Prenanthes muralis.
Reseda lutea; Luteola.
Sanicula europæa.
Silene inflata.
Solidago Virgaurea.
Spergula arvensis.
Spergularia rubra.
Stachys sylvatica.
Tanacetum vulgare.
Torilis Anthriscus.
Trifolium arvense.
Tussilago Farfara.
Verbena officinalis.
Vicia Cracca; sepium; tetrasperma.
Etc., etc.

SHOOTER'S HILL COMMON.

Castle Wood is enclosed; but the common fronting Blackheath is still open to the public. There are no trees: scrub or brushwood only, of Birch, Hazel, Oak, Sallows, White-thorn, Black-thorn, Briars, Furze, Broom, and Bramble. Subsoil, gravel.

*Achillea Ptarmica.
*Aira flexuosa; *cæspitosa.
*Betonica officinalis.
*Carex hirta; *ovalis.
*Centaurea nigra.
*Hieracium vulgatum; *umbellatum.
*Lotus major; *corniculatus.
*Ornithopus perpusillus.
*Potentilla Tormentilla; *Fragariastrum *repens, etc.
*Senecio sylvaticus.
*Solidago Virgaurea.
*Teucrium Scorodonia.

11. WANDSWORTH COMMON.

An extensive open space, of nearly three miles in circumference, partly bordered with villas, &c., and traversed by the Crystal Palace Railway, which has exercised upon it the ordinary effects of drainage. It is more or less covered with Furze, Bracken, and Ling; some parts are in fact densely furze-grown; others entirely free from it. There are several small ponds and pits, whence gravel has been excavated; and the subsoil is of this material. Nothing out of the common is to be found here now, but formerly its speciality was Stratiotes aloides.

*Acetosa Acetosella.
*Achillea Ptarmica; Millefolium.
*Aira præcox.
*Alisma Plantago.

*Alopecurus fulvus.
*Anthemis nobilis.
*Anthoxanthum odoratum.
*Betonica officinalis.
*Bidens tripartitus.
*Campanula rotundifolia.
Camelina sativa (near New Wandsworth station).
*Cardamine pratensis.
*Carduus acaulis; *pratensis.
*Carex flava; *hirta; *glauca; *ovalis.
*Centaurea nigra.
*Cerastium semidecandrum.
*Convolvulus arvensis.
*Elodea canadensis.
*Erigeron canadensis.
*Euphrasia officinalis.
*Filago minima.
*Galium saxatile; *uliginosum; *palustre.
*Geranium molle; *Robertianum.
*Glyceria fluitans.
*Gnaphalium uliginosum.
*Helosciadium inundatum.
*Hieracium Pilosella.
*Hydrocotyle vulgaris.
*Hypericum humifusum.
*Hypochœris radicata.
*Juncus effusus; *glomeratus; *supinus.
*Leontodon hirtus; *hispidus; *autumnalis.
*Lotus corniculatus.
*Lychnis diurna.
*Medicago maculata.
Mentha Pulegium; *hirsuta.
*Mœnchia erecta.
*Montia fontana.
*Myosotis arvensis; palustris.
*Nardus stricta.
*Nasturtium officinale; terrestre.
*Ononis spinosa.
*Ornithopus perpusillus.
*Pedicularis sylvatica.
*Peplis Portula.
*Polygonum Hydropiper, *Persicaria.
*Potamogeton natans; *crispum; *pusillum.
*Potentilla Tormentilla; *reptans; *anserina.
*Prunella vulgaris.
*Ranunculus aquatilis (agg.); *Flammula; *hederacea; sceleratus; acris; *bulbosus.
*Sagina procumbens.
*Salix repens.
*Scabiosa succisa.
*Scirpus palustris; *setaceus; *fluitans.
*Senebiera Coronopus.
*Senecio Jacobæa.
*Trifolium repens; *filiforme.
*Veronica hederifolia; *arvensis; *serpyllifolia; *Chamædrys; *Beccabunga.

CRYPTOGAMS.

*Bryum erythrocarpum.
*Ceratodon purpureus.
*Equisetum limosum; *palustre; *arvense.
*Pteris aquilina.

Clapham Common is furzy in some parts, and has some fine groups of trees upon it; it is surrounded by houses. Limosella aquatica grew formerly on the margins of the central pond; may be there still. Tooting Common is two miles in circumference, and furze-grown in parts, with Briars and a few Oak-trees at the further extremity. *Dicranum scoparium; *Polytrichum juniperinum; *Cratægus Oxyacantha, &c. Streatham Common is of the same character, but less extensive. *Anthemis nobilis; *Trifolium subterraneum; *Arenaria trinervis; *Achillea Ptarmica, &c.; on tombstones in Streatham churchyard, *Hypnum confertum; *serpens. They have both a gravelly subsoil. About Clapham Junction, *Erysimum cheiranthoides (abundant). Near the station, Wandsworth, Isatis tinctoria has been lately observed. Streatham Common is separated from that of Tooting by an avenue of Elms. Its subsoil is the same, and elevated at the upper end, where there is some Furze and Bramble, over three hundred feet. Senecio viscosus grew here formerly; and in the closes, about, Bunium flexuosum; Lactuca virosa; Daphne Laureola, &c.

12. ROADSIDES, COPSES, AND WASTE PLACES ABOUT NORWOOD.

The low hills which constitute the Southern Heights of London attain their highest elevation at Norwood, in connection with which are Gipsy Hill, Sydenham Hill and Forest Hill. Norwood is under four hundred feet high, less by about forty feet than Hampstead Heath at Jack Straw's Castle. Subsoil gravel. This portion of suburban London was unenclosed and unbuilt upon in Mr. Cooper's time; consequently little other than ordinary wayside plants are to be met with.

Æthusa Cynapium.
Agrimonia Eupatorium.
Ajuga reptans.
Bartsia Odontites.
Centaurea Jacea.
Chenopodium album; hybridum.
Chrysanthemum Leucanthemum.
Crepis virens.
Daucus Carota.
Epilobium montanum.
Galium Aparine; Mollugo; verum; saxatile.
Geranium molle; pusillum; dissectum; Robertianum.
Geum urbanum.
Hypochœris radicata.
Lactuca virosa.
Lapsana communis.
Lathyrus pratensis.
Leontodon hispidus; autumnalis.
Linaria vulgaris.
Lotus corniculatus.
Luzula pilosa; campestris.
Lychnis diurna.
Matricaria Chamomilla; inodora.
Mercurialis perennis.
Prunella vulgaris.
Ranunculus bulbosus; acris; repens.
Rosa canina.
Rubus fruticosus.
Rumex Acetosella.
Senecio Jacobæa.
Sinapis arvensis.
Solanum nigrum; Dulcamara.
Stellaria Holostea; graminea.
Tanacetum vulgare.
Taraxacum officinale.
Tragopogon pratense.
Tussilago Farfara.
Veronica; Vicia; Poa; Festuca; Bromus; Dactylis; Phleum, &c., ordinary species.

Copses.

Anemone nemorosa.
Bunium flexuosum.
Corylus Avellana.
Digitalis purpurea.
Euphorbia amygdaloides.
Hieracium vulgatum.
Ilex aquifolia.
Ligustrum vulgare.
Lysimachia Nummularia.
Melampyrum pratense.
Orchis mascula; maculata.
Orobus tuberosus.
Oxalis Acetosella.
Primula vulgaris.
Pyrus Aucuparia.
Ranunculus auricomus.
Ruscus aculeatus.
Sanicula europæa.
Scilla nutans.
Solidago Virgaurea.
Tamus communis.

Note.—Field, right of road from Croydon to Beulah Spa, ascending the hill: Sambucus Ebulus.

13. MITCHAM COMMON.

A gravelly waste, for the most part extending for a considerable distance in the direction of Croydon, and bordered in part by furze-grown patches. It is a level plain with many pits and hollows, where the gravel has been extracted; in many places filled with water. Mitcham Junction station is at the upper end, and a road to Croydon crosses the common from N.W. to S.E. West of it is the Wandle river, from which it is separated by some cultivated land and the road to Carshalton, &c. Mentha piperita is grown hereabout, and may be met with as a waif of cultivation.

*Actinocarpus Damasonium (E. de C.).
*Alchemilla arvensis.
*Alisma Plantago.
*Anthemis nobilis.
*Betonica officinalis.
*Bidens cernua; *tripartita.
*Calluna vulgaris.
*Clematis Vitalba (hedges near Mitcham).
*Cnicus acaulis.
*Euphrasia officinalis.
*Filago minima.
*Genista anglica.

*Geranium pyrenaicum.
*Hydrocotyle vulgaris.
*Hypericum humifusum.
Lathyrus Nissolia (at Mitcham) ? olim.
Leontodon palustre ; ? olim.
*Mentha Pulegium.
*Mœnchia erecta.
*Molinia cærulea.
*Myosotis versicolor.
*Nardus stricta.
*Ononis spinosa.
*Potamogeton densus (ponds).
Ranunculus parviflorus; ? olim.
*Sagina subulata.
*Senebiera Coronopus.
*Thymus vulgaris.
*Trifolium filiforme ; *subterraneum ; *striatum.
*Triodia decumbens.

CRYPTOGAMS.

*Chara vulgaris.
*Pilularia globulifera (E. de C.).

APPENDIX.

*Blysmus compressus (marshy patch, now enclosed, opposite Beddington Park gates).
*Phragmites communis (roadside towards Wimbledon) and *Medicago sativa (in fields, roadside towards Wimbledon).
*Scandix Pecten-Veneris (railway bank, Carshalton station).
*Sclerochloa rigida (by the park palings towards Wallington).
Scrophularia vernalis (grew formerly between Mitcham and Merton).

14. BANKS OF THE THAMES, WITH BORDERING DITCH AND MEADOWS BETWEEN KEW AND KINGSTON, RICHMOND HILL AND HAM COMMON.

A low-lying district, the hills excepted, with a gravelly subsoil. Ham Common is an open green with an avenue of Elms right of the high-road, wild and furze-grown left of it, and extending for nearly a mile to the borders of Richmond Park. A foot and towing path leads along by the river-side.

*Ægopodium Podagraria (about Richmond).
*Alisma Plantago (river-side).
*Anthemis nobilis (on the common).
Arnoseris pusilla (sandpit at Petersham ?).
*Butomus umbellatus.
*Calamintha Acinos (opposite Teddington).
*Campanula rotundifolia (Ham Common).
*Carduus crispus, &c.
*Carex riparia; *vulpina ; *paludosa.
*Centaurea nigra ; *Scabiosa (between Teddington and Kingston).
*Cerastium arvense (cornfields, Ham).
*Cichorium Intybus (opposite Teddington).
*Circæa lutetiana (below the hill).
*Clematis Vitalba (on the slope).
*Dianthus deltoides (gravelly pasture beyond Ham House).
*Digraphis arundinacea.
*Echium vulgare (opposite Teddington).
*Epilobium hirsutum.
*Fumaria officinalis (cornfields, Ham).
*Glyceria aquatica.
*Gnaphalium uliginosum (common).
*Hypericum humifusum (common).
*Jasione montana (common).
*Lamium amplexicaule (cornfields, Ham).
*Limnanthemum nymphæoides (in the river by Kingston, near the island).
*Linaria vulgaris.
*Lycopus europæus.
*Lythrum Salicaria.
*Nasturtium sylvestre ; *amphibium.
*Nepeta Cataria (hedge near Richmond).
*Nuphar lutea.
*Potamogeton perfoliatus.
Salix rubra (ditch between Kew and Richmond) ; *alba ; undulata ; *triandra ; fragilis ; viminalis.
*Salvia verbenaca (opposite Teddington ; scarce).
Saxifraga granulata (meadows, Richmond).
*Scabiosa succisa ; *arvensis.
Scilla autumnalis (meadows above Richmond, about Ham).
*Scutellaria galericulata.
*Sedum acre.
*Spiræa Ulmaria.
*Stachys palustris.
*Symphytum officinale (purple-flowered).
*Thymus Serpyllum (opposite Teddington).
*Torilis nodosa (opposite Teddington).
*Trifolium subterraneum (Ham Common) ; *arvense (cornfields) ; striatum.
*Typha latifolia (island near Kingston).
*Valeriana officinalis (scarce).
*Vicia Cracca.
*Viola arvensis (cornfields).

15. ROADSIDES ABOUT ISLEWORTH, TWICKENHAM, TEDDINGTON, HOUNSLOW, AND BANKS OF THE CRAN.

Low-lying and well-cultivated districts, subsoil, gravelly; but generally well covered with alluvium. Hounslow Heath, upon which many rare plants grew formerly, is now nearly all enclosed and converted into a military parade and rifle ground; a patch or two near the Cran is all that remains of it.

*Arabis Thaliana (bank, lower road, Teddington).
Arnoseris pusilla (gravelly fields, Teddington, olim, and fields near Hampton Court ?).
*Betonica officinalis (field near Cranford Bridge).
*Cardamine amara (Hanworth Bridge, abundant).
*Carex riparia; *paludosa; *hirta (Hanworth Bridge).
Dianthus prolifer (between Teddington and Hampton Court).
*Draba vernalis (walls Isleworth, and bank, lower road, Teddington).
*Erodium cicutarium (walls Isleworth, and bank, lower road, Teddington).
*Genista anglica (between Hospital and Hanworth bridges).
*Geranium molle; *pusillus (walls Isleworth, and bank, lower road, Teddington).
*Hottonia palustris (in the Cran at Hanworth Bridge).
*Hydrocharis Morsus-ranæ (in the Cran at Hanworth Bridge).
*Lemna trisulca (Hanworth Bridge).
Littorella lacustris ?
*Lysimachia vulgaris (Thames, Teddington).
Lythrum hyssopifolium ? (Hounslow Heath, formerly; the marshy corners near the bridges may be searched for this little plant).
*Myosotis versicolor (banks, roadside).
*Nuphar lutea.
*Papaver dubium (walls, Isleworth).
*Potamogeton pusillus; *lucens (Thames); *natans (Hanworth Bridge).
Rubus rhamnifolius; Kœhleri; Lindleianus; glandulosus.
*Salix purpurea (Mother Ives Bridge; Thames at Twickenham; *Hospital Bridge; Hanworth Bridge); *fragilis, &c.
*Saxifraga tridactylites (walls, Sion House).
*Scabiosa succisa (meadow near Cranford Bridge).
*Sedum acre (lower road, Teddington); album (walls between Isleworth and Brentford).
*Sium angustifolium (Cranford Bridge).
Teesdalia nudicaulis (Hampton Park).
*Typha latifolia; *angustifolia (Cranford Bridge).
*Valerianella olitoria (roadsides).
*Vicia lathyroides (roadsides).

CRYPTOGAMS.

Asplenium Ruta-muraria (Teddington church).
Confervæ (in the Cran).
*Homalothecium sericeum (walls at Bushy Park gateway).
Leskea polycarpa (trunks of Willows, Chiswick).
*Thuidium tamariscinum, with other common Hypna (roadside between Hospital and Hanworth bridges).

16. PINNER AND OXHEY WOODS, MEADOWS ABOUT PINNER AND RUISLIP, RUISLIP RESERVOIR.

About Pinner the country assumes a more hilly aspect than that which obtains immediately west of the metropolis, Harrow Hill being merely an isolated elevation in advance of the ridge which separates the valley of the Colne from that of the Brent. In the low-lying meadows hereabouts and at Ruislip the substratum is clay, but chalk underlies the gravel upon the hills beyond. Ruislip reservoir is a considerable piece of water, with reedy margins, and a wood on its eastern borders.

Alchemilla vulgaris (fields between the reservoir and Hareford road).
*Anemone nemorosa.
*Asperula odorata (Oxhey Wood).
*Briza media (borders Pinner Wood).
*Bunium flexuosum (borders Pinner Wood, plentiful).
*Carduus pratensis (Ruislip Moor).
*Carex glauca (Pinner and Pinner Hill); *curta; muricata (wayside hedges); *panicea; *pendula (brookside at Eastcot; also in Pinner Wood); *vulgaris (Ruislip Moor); *sylvatica (Oxhey Wood); *pallescens (Oxhey Wood); strigosa (Moss Lane, Pinner); disticha (reservoir).
*Conium maculatum (Ruislip Moor).
Epilobium roseum (Moss Lane).
Euphorbia amygdaloides.
*Fragaria vesca.
*Fritillaria Meleagris (Pinner, right of road to Rickmansworth near the town; also meadows, Ruislip?) olim.
*Hypericum perforatum; *hirsutum (woods).
*Lamium Galeobdolon (Pinner Wood).
*Leontodon hispidus.
*Lepidium campestre (Pinner Hill).
*Linum catharticum (Pinner Hill, pits &c.).
*Listera ovata (Pinner and Oxhey woods).
Littorella lacustris (Ruislip reservoir).
*Luzula sylvatica; *pilosa (woods); Forsteri.
*Lysimachia nemoralis (Pinner Wood); *Nummularia (Pinner Wood and Ruislip meadows).
*Melampyrum pratense (Ruislip woods).
*Melica uniflora (hedges).
*Myosotis sylvatica (Pinner and Oxhey woods); *arvensis.
*Narcissus Pseudo-Narcissus (Pinner).
Neottia Nidus-avis (Oxhey Wood).
*Orchis Morio (Ruislip Moor and Pinner); maculata; *mascula (woods); *latifolia (moor).
*Populus alba (Pinner Wood).
*Primula veris (meadows); *vulgaris (woods).
*Ranunculus auricomus.
*Rhinanthus Crista-Galli (meadows).
*Salix cinerea; Caprea.
*Scilla nutans.
*Sedum reflexum (Pinner, roadside; garden waif).
*Stellaria graminea.
*Tamus europæus.
*Valeriana dioica (Ruislip Moor).

CRYPTOGAMS.

*Dicranum scoparium (Pinner).
*Equisetum palustre (Ruislip Moor).
Hypnum striatum; Schreberi (Pinner).
Jungermannia asplenioides; heterophylla; complanata (Pinner).
Mnium undulatum (Pinner).
*Nephrodium Filix-mas (woods).
Orthotrichum affine (Pinner); crispum.
Tortula fallax (Ruislip); subulata (Pinner).

17. HARROW WEALD COMMON, STANMORE HEATH, ELSTREE RESERVOIR.

Stanmore Heath, three miles from the village of Edgeware, is five hundred feet high; elevated on the north, west and south; with a fall on these slopes to a depression on its eastern edge, where there is a large pond fed by the converging lines of drainage. This pond is in private grounds, and connected with a larger one near Elstree, known as Elstree reservoir. The road to Watford skirts the western borders of the heath, and divides it from the grounds of Bentley Priory; due west, and distant a mile or so, is Harrow Weald Common; on the same line of elevation, north and west of the Priory. The Heath is turfy and furze-grown, except on the lines of drainage, which are open and marshy; subsoil gravelly, and pitted near the village with shallow excavations, usually full of water. At the southern extremity is a grass plot. The Common is a narrow strip of wet gravelly waste, about three-quarters of a mile long, with an outlying copse or two in the directiou of Pinner; it slopes to the north-west; without turf and of a less furzy character than the Heath, than which it is a trifle less elevated.

Stanmore Heath.

*Achillea Ptarmica.
*Alopecurus geniculatus.
*Anthemis nobilis.
*Asperula odorata (between Elstree and Stanmore lower road).
*Calluna vulgaris.
*Carex flava; lævigata; *panicea; *Pseudo-cyperus (pond in a field near Whitchurch).
Epilobium obscurum.
*Erica Tetralix.
*Genista anglica.
Hieracium boreale ?
*Juncus squarrosus, &c.
Linum catharticum.
*Orchis maculata.
*Prunus spinosa.
Pyrola minor (under some old trees east of the heath).
Rubus fruticosus; agg.
*Salix repens and var.
Sanguisorba officinalis (pastures near) ?
*Scilla nutans (hedgerows, Whitchurch).
*Senecio erucifolius.

Cryptogams.

*Aulocomnion palustre.
*Bartramia pomiformis (bank left of road from Elstree station to Stanmore).
Chara translucens.
*Homalothecium sericeum (wall opposite Whitchurch Church).
*Hypnum albicans; splendens; *cuspidatum.
Nephrodium dilatatum.
*Pteris aquilina.
*Thuidium tamariscinum.
*Weissia cirrhata (palings Stanmore), &c.

Harrow Weald Common.

Besides Furze, Heath, Briars, Brambles, and Bracken; the following :—

*Digitalis purpurea.
Drosera intermedia.
Epilobium obscurum.
*Erica cinerea.
Gnaphalium sylvaticum.
Hieracium boreale ? *umbellatum.
Luzula Fosteri (copses near).
Populus tremula.
Sagina ciliata.
*Salix cinerea; *repens.
*Scutellaria minor.
*Teucrium Scorodonia.
Vinca minor (woods, Bentley Priory).

Cryptogams.

Asplenium Adiantum-nigrum.
Athyrium Filix-fœmina.
*Atrichum undulatum.
*Aulocomnion palustre.
*Carex ovalis; *stellulata; *panicea; *glauca.
*Dicranella heteromalla; *varia.
*Dicranum scoparium.
*Equisetum sylvestre (lower border by the hedge).
*Fissidens taxifolia; *bryoides (roadside ditch-banks, towards Harrow).
*Fontinalis antipyretica (field, left towards Kenton Lane).
*Hypnum cupressiforme; *purum; *cuspidatum; *splendens; *squarrosum; *stellatum; glareosum; *cordifolium; *sylvaticum (bordering shady banks); myuroides; *triquetrum; loreum ?; *aduncum.
Isothecium alopecurum.
Jungermannia undulata; *scalaris; *Trichomanes; *crenulata; *dilatata; *calycina.
Leskea polycarpa.
Leucodon sciuroides.
*Leucobryum glaucum.
*Lomaria Spicant.
*Mnium undulatum; *hornum; *rostratum.
*Neckera complanata (Kenton Lane).
Orthotrichum affine.
*Pleuridium subulatum.
*Pogonatum aloides.
*Polytrichum commune; *juniperinum; *piliferum.
*Sphagnum cuspidatum.
*Tortula unguiculata.

Elstree Reservoir.

Carex vesicaria.
Limosella aquatica.
Scirpus acicularis.
Tortula subulata.

18. TOTTERIDGE GREEN AND HADLEY COMMON.

The ridge referred to in the previous section is continued eastwards towards High Barnet, which stands on ground of the same elevation as Stanmore Heath; beyond Barnet is Monken Hadley and a wood known as

Hadley Common. Mill Hill and Totteridge are separated from Barnet by a valley in which runs the Brent stream. This stream encloses Totteridge heights in a loop, and then runs westward towards Hendon. Subsoil gravel.

TOTTERIDGE GREEN.

*Acorus Calamus (ponds).
*Carex divulsa (hedgeside in a field on the left by footpath from lower end of the green to the Brent).
Chara flexilis (ponds).
*Crocus vernus (meadows near).
Dianthus deltoides (back of Osmund's barn).
Fritillaria Meleagris (meadows near).
*Hottonia palustris (ponds).
Lilium Martagon (Totteridge Park).
Mentha Pulegium.
*Ranunculus Lingua (ponds).

HOLLOW BELOW; BY THE BRENT.

*Allium ursinum.
*Alnus glutinosa.
Equisetum maximum ? (between Totteridge and Barnet).
*Rhamnus catharticus (opposite Woodside station).
*Rhinanthus Crista-Galli (meadows).
*Scirpus sylvaticus.
Spiranthes autumnalis (also on Mill Hill).
*Viburnum Opulus.
Vinca minor (near Totteridge).

HADLEY COMMON.

A wood sloping from west to east; where the ridge joins that which borders the vale of the Lea westwards. Oak, &c., with an underscrub of White-thorn, Briar, and Holly. Open grass glades in parts.

*Bunium flexuosum.
*Carex sylvatica.
*Daphne Laureola.
*Euphorbia amygdaloides.
*Fragaria vesca.
*Lysimachia nemorum.
*Melica uniflora.
*Oxalis Acetosella.
*Primula vulgaris.
*Sanicula europæa.
*Scilla nutans.

19. EPPING UPPER FOREST.

From Buckhurst Hill to Epping town, a road—viâ High Beech green—traverses the upper forest for its whole length in a north-easterly direction for a distance of six miles. Forest land thickly wooded on the flanks, and more or less so on the plateau above, where, however, there are occasional more open patches of moor and heath, with frequent pits and pools. Near High Beech and the Royal Oak are some small boggy bits, and within a mile or two of Epping an ancient entrenchment called Amesbury Banks; between this and the Royal Oak is a central station, the Wake Arms, where cross roads meet from Waltham, Loughton, and Theydon Bois; the best localities for plants, especially Ferns, are the gullies and hollows on the forest flanks; others must be sought for in the bordering fields near Epping; there is no abundance of anything out of the common in the forest itself, however productive it may have been formerly. On the south-east flank the lines of drainage are to the Roding, on the north-west to an affluent of the Lea, called Cobbins Brook; the vale traversed by this stream divides the forest upland from Nasingwood Common and adjacent high ground to the eastward of Cheshunt and Broxbourne. The subsoil is gravel; and the rise very gradual. The general vegetation is Hornbeam and Beech, both much lopped, Oak scrub, Holly, Bramble, Briar, Bracken,

White and Black-thorn, Lonicera, Crab-apple, Furze, and a little Birch, Sallows and Salix repens, Ling and Rushes.

*Acer campestre.
*Achillea Ptarmica.
*Aira flexuosa; *præcox; *cæspitosa.
Anagallis tenella (bog, High Beech).
*Anemone nemorosa.
Arabis perfoliata.
Asperula odorata.
*Betonica officinalis.
*Bunium flexuosum.
Calamagrostis Epigejos.
Camelina sativa (field near Epping).
Carex pilulifera; binervis; *ovalis; *flava; remota; *stellulata; *panicea; vulgaris.
Centunculus minimus (near High Beech).
Chrysosplenium oppositifolium; alternifolium.
*Circæa lutetiana (Amesbury banks).
*Cornus sanguinea.
*Corylus Avellana.
Cynoglossum montanum.
*Daphne Laureola.
Dianthus Armeria (towards Theydon).
*Digitalis purpurea.
Dipsacus pilosus.
Drosera rotundifolia (scarce ?).
*Epilobium montanum; roseum.
*Erica cinerea; Tetralix (scarce).
*Erythræa Centaurium.
*Euphorbia amygdaloides.
*Euphrasia officinalis.
Festuca Pseudo-myurus (dry parts).
*Fragaria vesca.
*Fraxinus excelsior.
Fumaria officinalis (fields near).
*Galium Mollugo; verum; *palustre.
*Genista anglica.
*Geranium dissectum, &c.
Gnaphalium sylvaticum.
*Hedera Helix.
*Hieracium vulgatum.
*Holcus mollis.
*Hydrocotyle vulgaris.
*Hypericum pulchrum; *humifusum; Androsæmum.
Isatis tinctoria (near Epping, in fields).
*Juncus glomeratus *squarrosus, &c.
*Lamium Galeobdolon.
*Lathyrus pratensis.
*Ligustrum vulgare.
Listera ovata.
*Lotus major; corniculatus.
*Lychnis diurna; *Flos-cuculi.
Lysimachia nemorum; Nummularia.
*Melampyrum pratense.
*Mentha sativa (Amesbury banks).
Mœnchia erecta.
*Molinia cærulea.
Myosurus minimus.
*Myriophyllum verticillatum (ponds).
*Nardus stricta.
*Orchis maculata; mascula.
*Ornithopus perpusillus.
*Orobus tuberosus.
*Oxalis Acetosella.
Parnassia palustris ? (nearer than Ongar ?)
*Pedicularis sylvatica; palustris.
*Polygala vulgaris.
Populus alba; tremula.
Potentilla argentea; *Fragariastrum, &c.
*Primula vulgaris.
Pyrus Aucuparia.
*Ranunculus auricomus; *Flammula.
*Rubus (several species; see Index).
Ruscus aculeatus.
*Salix Caprea *repens, and var.
*Sambucus Ebulus (near Sewardstone, towards Waltham).
*Sambucus nigra.
*Sanicula europæa.
*Sarothamnus Scoparium.
*Scabiosa succisa.
*Scilla nutans.
*Scrophularia Balbisii; *nodosa.
Senebiera didyma; *Coronopus.
*Senecio sylvaticus; erucifolius; *Jacobæa.
Serratula tinctoria.
*Solidago Virgaurea.
*Stachys sylvatica.
*Tamus communis.
Teesdalia nudicaulis.
*Teucrium Scorodonia.
*Tilia europæa.
*Trifolium procumbens; *filiforme.
*Triodia decumbens.
*Typha angustifolia (pond).
*Ulex Gallii.
*Ulmus suberosa; montana.
*Veronica officinalis; montana; scutellata; *Chamædrys.
*Viburnum Opulus.
Viscum album.
Wahlenbergia hederacea (?) (r. of Abridge road).

CRYPTOGAMS.

*Aspidium aculeatum, and var. lobatum; angulare.
Asplenium Ruta-muraria.
Ceterach officinarum.
*Chara flexilis; translucens.
Equisetum sylvaticum.
Lomaria Spicant.
Lycopodium clavatum (?); olim (near High Beech).

*Nephrodium Filix-mas; spinulosum; dilatatum; Thelypteris; Oreopteris.
*Polypodium vulgare.
*Pteris aquilina.
Scolopendrium vulgare.

*Atrichum undulatum.
*Dicranella cerviculata; *heteromalla.
*Dicranum scoparium.
*Fungi and Lichens (many).
*Hypnum splendens; *triquetrum; &c.
*Jungermanniæ (several).
*Leucobryum glaucum (near Wake Arms in a beech copse).
*Polytricha (three species).
*Sphagnum (near High Beech).
*Thuidium tamariscinum.
*Weissia controversa, &c.

The copses and woodlands beyond Epping, several detached pieces, are private property and enclosed. In a broad ditch, border of upper forest towards Roydon, Peucedanum officinale?

20. HAINAULT FOREST AND BANKS OF THE RODING.

Although Hainault Forest, once very extensive, no longer exists but in name, some detached bits of scrub and underwood about Chigwell Row are still standing, and these, together with Crab-tree Wood, may be searched for what may yet be obtained. East of this, and on the other side of the Bourne brook, is Havering atte Bower, where are the remaining walls of an ancient palace. The Roding is a sluggish stream, meandering through bordering meadows along the vale which lies between the Hainault and the Epping uplands. Aquatic plants grow here in plenty.

Copses, etc., Chigwell Row.

Calamagrostis Epigejos.
Cynoglossum montanum.
Hypericum pulchrum; Androsæmum.
Lysimachia nemorum.
Ruscus aculeatus.
*Sambucus Ebulus (lane near Chigwell church, leading to the Roding).
Vinca minor (Theydon Bois).

Cryptogams.

Aspidium aculeatum (about Chigwell).
Nephrodium Thelypteris (about Chigwell).

Banks of the Roding.

Acorus Calamus.
*Alnus glutinosa.
*Carex acuta; *riparia; *vesicaria.
*Digraphis arundinacea.
*Epilobium hirsutum.
*Helosciadium nodosum.
*Iris Pseudacorus.
Limnanthemum nymphæoides.
*Lythrum Salicaria.
*Myosotis palustris.
*Nuphar lutea.
Œnanthe fluviatilis (Woodford).
*Polygonum amphibium.
*Potamogeton natans; *densus; *perfoliatus; *lucens.
*Rumex Hydrolapathum.
*Sagittaria sagittifolia.
*Salix alba; cinerea; fragilis.
*Scirpus lacustris.
*Scutellaria galericulata.
*Sium angustifolium; latifolium (near Wanstead).
*Sparganium ramosum.
*Spiræa Ulmaria.
*Stachys palustris.
*Stellaria aquatica.
*Symphytum officinale.
*Thalictrum flavum.
Typha angustifolia.
*Viburnum Opulus.

21. MARSHES BETWEEN WOOLWICH, PLUMSTEAD, AND ERITH, AND OPPOSITE SHORE.

The flats below Woolwich towards Erith, formerly extensive marshes, have been drained and converted into pasturage. To this end the embankment by the river-side has materially contributed; and consequently aquatic plants, both rare and ordinary, have in a great degree disappeared; what remains must be sought for on the margins of the ditches, wherever these may be accessible. To what extent the embankment has prejudiced the riverside vegetation is not easy to determine. Subsoil alluvial. The shores and flats on the opposite shore are of the same character.

*Apium graveolens.
*Artemisia maritima.
*Asparagus officinalis (two or three plants; on the other side, more perhaps).
*Aster Tripolium.
*Beta maritima (near Erith).
*Carex divisa; *vulpina.
Caucalis daucoides (?) (near Erith).
*Chenopodium Bonus-Henricus (Erith).
*Cochlearia anglica.
*Conium maculatum.
*Crepis paludosa.
*Cynoglossum officinale (ditch banks towards Erith; a few plants).
*Festuca sciuroides (pastures).
*Fœniculum vulgare (a few plants).
*Glaux maritima.
*Helminthia echioides.
*Hydrocharis Morsus-ranæ.
*Juncus acutiflorus; *glomeratus, &c.
*Lactuca saligna (near Erith).
*Lepidium ruderale.
*Lepturus filiformis.
*Medicago maculata; sativa.
*Phragmites communis.
*Plantago maritima.
Polypogon monspeliensis (near the rifle butts, scarce. W. Reeves.)
*Potamogeton pusillus; *pectinatus.
*Samolus Valerandi.
*Scirpus maritimus.
*Sclerochloa maritima.
*Sinapis nigra.
*Spergularia neglecta (marina seg.).
*Suæda martima.
*Torilis nodosa (banks, plenty).
*Tragopogon pratensis; porrifolius (?) olim.
*Trifolium maritimum (?) (not seen).
*Triglochin palustre; *maritimum.
*Typha latifolia.
*Vicia tetrasperma.
*Zannichellia pedicellata.

Sonchus palustris, Leucojum æstivum, Polypogon littoralis, grew here formerly. The first-mentioned two plants also in the marshes about Plaistow, on the opposite side of the river, and the other near the powder magazine, between Woolwich and Erith.

Agrostis Spica-Venti; between Ilford and Barking.

Dipsacus pilosus; between Wanstead and Barking.

Lepidium latifolium and Draba; about Barking.

Gastridium lendigerum, Sclerochloa loliacea, Hypochœris glabra, with Chenopodium olidum, grew formerly in the Woolwich warren; may be there now.

22. PLUMSTEAD COMMON, BOSTALL HEATH, ABBEY WOOD, ERITH SAND-PITS.

The heights which serve as boundary inland to the flats between Woolwich and Erith, are between two and three hundred feet high. Plumstead Common does not exceed a hundred and fifty, with a gradual rise. It is a gravelly plateau at the foot of Shooter's Hill, furzy and uneven in the direction of Bostall Heath, with which it is nearly continuous; here are sand-pits, and beyond is Abbey Wood; near it, towards the flats, are the ruins of Lesnes Abbey. The grounds of Belvedere Park come between the wood and the sand-pits at Erith.

ON THE COMMON.

*Aira flexuosa; *præcox.
*Leontodon hirtus; *hispidus.
*Senebiera Coronopus, &c.
*Spergularia rubra.

ABBEY WOOD.

Acer campestre.
Betula alba.
Carpinus Betulus.
Clematis Vitalba.

Euonymus europæus.
Fraxinus excelsior.
Fumaria capreolata (agg.).
Ruscus aculeatus.
Ulmus montana (?); suberosa and var. (woodland mosses—in plentiful variety).

ERITH SAND-PITS.

*Chenopodium polyspermum.
*Lathyrus sylvestris.
*Melilotus officinalis; alba.
*Papaver Rhœas; dubium.
*Pastinaca sativa.
*Reseda Luteola.
*Spergularia rubra.
*Trifolium procumbens; minus.
Etc., etc.

23. CHISLEHURST COMMON,

Three hundred feet above the sea level, consists of two portions; the one, turfy and furze-grown, bordered by the village and by the grounds of Camden House; the other, east of the old church, of a wilder and more varied character, surrounded by woods and crossed by the road to St. Mary Cray. The subsoil is gravel; and the drainage through wooded slopes to a brook in the hollow south of the plateau. Several plants of rare occurrence grew in the bogs and damp bottoms of these woods; may be there now, but the woods are all enclosed, with the exception of the public thoroughfares through them.

WEST COMMON.

Much Furze, with Fern and Heath, intersected by roads and footpaths, with pools, marshy and gravelly open patches on the borders.

*Alchemilla arvensis.
*Campanula rotundifolia.
*Draba verna.
*Mœnchia erecta.
*Peplis Portula.
*Scirpus palustris.
*Vicia sepium (banks).

EAST COMMON.

Heath and Ling interspersed with Bracken, Furze, Blackthorn, Brambles, and Birch; moory in parts; gravel-pits. In the woods, Oak and Birch, chiefly with the dense undergrowth of Bracken, Holly, and on the borders, Yew.

*Achillea Ptarmica.
*Allium ursinum (brook below).
*Alnus glutinosa (hollows below).
*Anagallis tenella.
*Anemone nemorosa.
*Betonica officinalis.
Blysmus compressus (bogs)?
*Campanula rotundifolia.
*Carex vulpina; *panicea; *sylvestris.
*Centunculus minimus.
*Circæa lutetiana.
*Clematis Vitalba.
Convallaria majalis (wood left near the village).
*Digitalis purpurea.
*Euonymus europæa.
*Euphorbia amygdaloides.
*Galium uliginosum.
*Hieracium umbellatum; *vulgatum.
*Hydrocotyle vulgaris.
*Hypericum hirsutum; *tetrapterum; *humifusum; Androsæmum (wood between Chislehurst and Bromley).
Impatiens Noli-me-tangere? (wood).
Lathræa Squamaria (wood S. of bog).
Lathyrus Nissolia (roadside near church).
*Lepidium Smithii.
*Lithospermum officinale (copse below wood S. of E. Common).
*Lysimachia Nummularia; *nemorum.
*Lythrum Salicaria.
*Melampyrum pratense.
*Mentha sativa; Pulegium; *hirsuta.
*Milium effusum.

*Nardus stricta.
*Oxalis Acetosella.
Paris quadrifolia (Petz bog, also in the Long Wood); olim.
Pinguicula vulgaris (?) (Petz bog); olim.
Ranunculus parviflorus.
*Rhamnus Frangula (hollow below).
*Ribes rubrum (wood between church and Bromley); Grossularia (wood between church and Bromley).
*Rosa arvensis.
*Rubus glandulosus; *carpinifolius;* plicatus (hollow below).
*Ruscus aculeatus.
*Salix repens; *Caprea; *cinerea.
Saxifraga granulata; *tridactylites.
*Scabiosa succisa.
*Scirpus setacea.
Scrophularia vernalis.
*Scutellaria minor.
*Solidago Virgaurea.
*Spiræa Ulmaria.
*Teucrium Scorodonia.
*Viburnum Opulus.

Wood between Chislehurst and Orpington.

Dipsacus pilosus.

Cryptogams.

Chara hispida; tomentosa.
Dicranella cerviculata.
Equisetum sylvaticum.
Hypnum filicinum, &c.
*Mnium undulatum (shady bank, roadside, near St. Mary Cray).

Note.—The country about the Crays is all enclosed or cultivated, and roadsides too well trimmed for wild plants to flourish.

24. HAYES AND KESTON COMMONS.

These commons are continuous one with the other. Their elevation is over three hundred feet above the bed of the Thames, with a gravelly subsoil. A mile or two beyond Keston, this gravel drift disappears, and the chalk which underlies it comes to the surface; there are indications of this approaching change in the character of the soil (in a southerly and easterly direction) by the occurrence, though sparingly, of two or three characteristic plants. Between the commons are some reservoirs of spring water.

Hayes Common

Lies to the right of the high-road from Bromley to Westerham, and is a fine open expanse, overgrown with a short dense scrub of Heath and Ling; Furze, Bracken and Broom; Black-thorn and White-thorn.

*Aira præcox; *flexuosa.
*Alchemilla arvensis.
*Campanula rotundifolia.
*Carex præcox.
*Cistus Helianthemum.
*Clematis Vitalba (roadside between the commons).
*Cuscuta Epithymum.
*Euphrasia officinalis.
Galium verum; cruciatum.
*Hypericum pulchrum; *humifusum.
*Linum catharticum.
*Lotus corniculatus.
*Marrubium vulgare (near the mill).
*Melampyrum pratense.
*Menyanthes trifoliata (bog).
*Ornithopus perpusillus.
*Oxalis Acetosella.
*Pimpinella Saxifraga.
*Potentilla Tormentilla.
*Rhamnus Frangula.
*Rumex Acetosella.
*Ruscus aculeatus.
*Spergula arvensis.
*Spergularia rubra.
*Teucrium Scorodonia.
*Thymus Serpyllum.
*Triodia decumbens.
*Viola sylvatica.
Wahlenbergia hederacea (about one of the reservoirs)?

Keston Common

Stands on higher ground and is higher at the upper than at the lower end, which adjoins the preceding; fronting the windmill is a hollow with boggy ground extending downwards, and on the other side of the slope beyond is a reservoir fed by a spring known as Cæsar's Well; this reservoir flows into another larger and enclosed one. The spring is in fact the source of the Ravensbourne rivulet. Bordering the common on one side is Holwood Park, where are extensive remains of an ancient entrenched town or camp. On the other side is a wood sloping to the south-west. The common is overgrown with Furze, Bracken, and Ling.

*Anagallis tenella.
*Briza media.
*Carduus pratensis.
*Carex flava; stellulata; *panicea.
*Digitalis purpurea (copse near).
*Drosera rotundifolia.
*Erica Tetralix.
*Euphrasia officinalis.
*Filago minima.
*Galium palustre.
*Genista anglica; tinctoria (?).
*Hydrocotyle vulgaris.
*Hypericum Elodes; *pulchrum; *tetrapterum.
Lepidium Smithii.
*Linum catharticum.
Listera ovata (bordering wood ?).
*Lychnis Flos-cuculi; *diurna.
*Lysimachia Nummularia; *nemorum (adjoining wood).
*Molinia cærulea.
*Nardus stricta.
*Narthecium ossifragum.
*Orchis maculata.
*Orobus tuberosus.
*Pedicularis sylvatica; *palustris.
*Polygala vulgaris.
*Potamogeton polygonifolius.
Radiola Millegrana.
*Ranunculus sceleratus.
*Rubus Idæus.
Scirpus setaceus; *fluitans; *multicaulis.
*Scutellaria minor; *galericulata.
*Sparganium ramosum.
*Thymus Serpyllum.
*Trifolium filiforme; arvense.
*Vaccinium Myrtillus.
*Veronica scutellata; *serpyllifolia; *Anagallis; * Beccabunga.

Cryptogams.

Asplenium Adiantum-nigrum (?).
*Bertramia fontana.
*Lomaria Spicant.
*Lycopodium inundatum.
*Nephrodium Filix-mas.
Polypodium vulgare.
*Sphagnum cuspidatum.

Appendix.

About Bromley: Arabis perfoliata (abundant formerly); Dianthus Armeria; *Potentilla argentea (sandy banks); Verbascum Lychnitis ? (seen at Bickley); *Thapsus (hedge on the common); on the chalk towards Down, Gentiana Amarella. From Down towards Cudham and Knockholt Beeches, other plants of the chalk formation will be met with; fields near, Camelina sativa.

25. SHIRLEY COMMON AND THE ADDINGTON HILLS.

South-east of Croydon, and over four hundred feet in height; western slope undulating and furrowed at intervals, with lines of drainage; level above, and bordered on the east by Addington Park palings; sand-pits at the upper extremity. Below, a pond and some wet places, a strip of copse, and a plantation beyond, in the direction of Croham Hurst. Subsoil, gravel. At the further extremity of the plantation, and facing this elevation, is a chalk-pit. The common and hills are covered with Heath; a cross road leads towards Shirley from the Addington road, through the copse at the foot of the slope.

*Alchemilla arvensis.
Blysmus compressus (bogs near).
Carex dioica; pulicaris (bogs near); *præcox; *binervis.
Centunculus minimus.
*Cerastium semidecandrum.
Drosera intermedia (bogs near).
Eriophorum vaginatum.
*Erodium cicutarium.
*Genista anglica.
Hypericum Elodes.
*Mœnchia erecta.
*Molinia cærulea (ravines).
*Myosotis versicolor.
Narthecium ossifragum.
*Polygala vulgaris.
Rhynchospora alba (bog).
*Saxifraga granulata (near the pond).
*Trifolium subterraneum.
Viola palustris (meadow near).

Note.—The bog is on the road to Wickham; on inquiry, could hear of no other on or about the common.

CRYPTOGAMS.

Bartramia fontana.
Botrychium Lunaria (?).
Lomaria Spicant.
Lycopodium Selago.

ADDENDA.

Chalk-pit referred to: *Linaria minor.
*Helianthemum vulgare; surrejanum (borders of a wood near? *olim*).
Gravel-pits and waste gravelly places about Croydon: *Medicago maculata, lupulina, falcata?; Dianthus Armeria; Malva moschata.
Duppas Hill: Dianthus deltoides; Silene anglica, quinquevulnera, *olim.*
Watery places near Croydon: Limosella aquatica.

26. CROHAM HURST AND THE ADJOINING FIELDS, BANKS, AND ROADSIDES.

Opposite the Addington Hills, from which it is separated by a narrow vale, is Croham Hurst, an isolated ridge, four or five hundred feet high of pebbly gravel, resting upon chalk and surrounded by bordering banks of this formation. It extends for nearly a mile in the direction of Selsdon and Sanderstead; furzy and heathy on the narrow ledge above, but well wooded on the flanks. The roads leading to the above-named localities are prolific in chalk plants, others grow in the adjoining fields. Altogether the flora of this locality is a very varied one, although, probably, no longer so abundant as may have been the case formerly.

Aceras anthropophora.
*Adoxa moschatellina.
*Ajuga Chamæpitys (fields).
Anchusa arvensis (fields).
*Anemone nemorosa.
*Anthyllis Vulneraria.
Aquilegia vulgaris.
Arenaria serpyllifolia; trinervis.
*Asperula cynanchica.
Astragalus Glycyphyllos.
*Betonica officinalis.
*Bunium flexuosum.
*Calamintha Clinopodium; *Acinos.
Campanula glomerata.
*Carduus acaulis.
*Centaurea Scabiosa; *nigra.
Chlora perfoliata.
*Circæa lutetiana.
*Clematis Vitalba.
*Convallaria majalis.

Cuscuta Epithymum.
*Daucus Carota.
*Digitalis purpurea.
*Echium vulgare.
*Erica cinerea; *Tetralix (sides of ridge).
Erigeron acris.
Epipactis latifolia (?).
*Euphorbia amygdaloides; exigua.
*Fumaria officinalis (fields).
Galanthus nivalis.
*Galeopsis Ladanum (fields near).
*Galium tricorne (fields); *verum.
*Genista anglica.
Gentiana Amarella.
Geranium columbinum.
*Gymnadenia conopsea (bordering banks).
Habenaria bifolia (?).
*Helianthemum vulgare.
*Hieracium vulgatum; *umbellatum.
*Hippocrepis comosa.
*Hypericum perforatum; *hirsutum.
Lamium Galeobdolon.
*Lepidium campestre (fields).
*Linaria spuria.
*Linum catharticum.
Listera ovata.
Lithospermum arvense (fields).
*Lotus corniculatus.
*Lysimachia nemorum.
*Melampyrum pratense.
*Mercurialis perennis; annua.
*Myosotis versicolor.
Narcissus Pseudo-Narcissus (farm).
*Nepeta Cataria (roadside).
Onobrychis sativa.
*Ononis arvensis.
Ophrys muscifera; apifera (?) (bordering banks).
*Orchis mascula; pyramidalis.
*Origanum vulgare.
*Orobanche minor (clover fields near).
*Orobus tuberosus.
*Phyteuma orbicularis (road to Sanderstead).
*Polygala vulgaris.
Potentilla argentea.
*Poterium Sanguisorba.
*Primula vulgaris.
*Prunus Cerasus (woods).
*Reseda lutea.
Rosa spinosissima; rubiginosa.
*Ruscus aculeatus.
*Sanicula europæa.
*Scabiosa arvensis; *Columbaria.
*Scilla nutans.
Sedum Telephium.
Serratula tinctoria.
*Sherardia arvensis.
*Silene inflata.
Specularia hybrida (fields).
Spiræa Filipendula.
Stachys arvensis (fields near).
*Tamus europæus.
*Taxus communis.
*Teucrium Scorodonia.
Thesium humifusum (?).
*Thymus Serpyllum.
Trifolium scabrum.
*Vaccinium Myrtillus.
Valerianella dentata (fields).
*Verbascum nigrum; Lychnitis (?).
*Viburnum Lantana; *Opulus.
Vinca minor (?).
Viola canina (flavicornis var.).

CRYPTOGAMS.

Polypodium vulgare.
Trichostomum canescens.

27. PURLEY DOWNS AND RIDDLESDOWN, SMITHAM BOTTOM.

Beyond Croham Hurst, and distant three miles from Croydon. Purley Downs are in part enclosed, and converted into a rifle practising ground direction north and south, and sloping westwards into a vale below beyond which rises the east flank of Riddlesdown. But as the direction of this chalk hill is south-east and north-west, a section of country intervenes which is of an undulating character, and thus the further end of Riddlesdown abuts upon the Brighton road at Caterham Junction, while its south-westerly flank bounds Caterham Valley to the north-eastward. Smooth grassy downs, with scattered Juniper in places; subsoil, chalk elevation about five hundred feet. At Smitham Bottom is a large chalk pit, right of the Brighton road. At Caterham Junction and at Riddlesdown are other and more extensive quarries.

Aceras anthropophora (?)
Ajuga Chamæpitys (fields near).
Anthyllis Vulneraria (pits and banks).
*Arabis hirsuta.
*Arenaria trinervis (bordering fields).
*Asperula cynanchica.

*Briza media.
*Calamintha Clinopodium; *Acinos.
*Carduus acaulis.
*Chlora perfoliata (about the pits).
*Clematis Vitalba (hedges near).
*Daucus Carota.
*Echium vulgare (chalk pits mostly).
*Galeopsis Ladanum (fields near).
Gentiana Amarella.
*Geranium Robertianum (pits).
Gymnadenia conopsea.
*Habenaria viridis.
*Helianthemum vulgare.
Herminium Monorchis.
*Hippocrepis comosa (chalk pits).
*Hypericum hirsutum; *montanum (copse near Caterham Junction); *perforatum.
*Jasione montana.
*Juniperus communis.
*Linaria minor (pits and fields).
*Linum catharticum.
*Nepeta Cataria (Smitham Bottom).
*Onobrychis sativa.
*Ononis arvensis.
*Onopordum Acanthium (Smitham Bottom, a plant or two).
Ophrys muscifera (?); apifera (?).
Orchis pyramidalis.
*Origanum vulgare.
*Parietaria diffusa.
*Phyteuma orbicularis.
*Plantago media.
*Poterium Sanguisorba.
*Reseda lutea; *Luteola.
*Rosa spinosissima; *rubiginosa; *micrantha (copses, Riddlesdown).
*Scabiosa Columbaria; *arvensis.
*Sclerochloa rigida (bordering fields).
*Spiræa Filipendula.
Spiranthes autumnalis.
*Thymus Serpyllum.
*Verbascum nigrum (Smitham Bottom, also chalk-pits); *Thapsus (pits); *Lychnitis (pit at Riddlesdown—plentiful).
Verbascum nigrum (about the railway-station, Caterham Junction).
*Viburnum Lantana.

28. FARDEN (OR FARTHING) DOWNS; WITH COULSDON AND KENLEY COMMON.

Two miles beyond Caterham Junction, and left of the Brighton road are Farthing Downs, and separated from the high ground about Coulsdon by a valley running from north-west to south-east, towards Upper Caterham. On either side of the road, the chalky slopes are cultivated; right of it, however, are two small unenclosed patches, with Juniper and other scrub upon them. Here—

*Anthyllis Vulneraria.
*Asperula cynanchica.
*Briza media.
*Carduus acaulis.
*Carlina vulgaris.
*Helianthemum vulgare.
*Linum catharticum.
*Onobrychis sativa.
*Reseda lutea.
*Rosa micrantha.
*Scabiosa Columbaria.
*Spiræa Filipendula.
*Thymus Serpyllum.

BORDERS OF FIELDS ABOUT THESE PATCHES.

*Ajuga Chamæpitys.
*Calamintha Acinos; Clinopodium.
*Carduus nutans.
*Crepis taraxacifolia.
*Galeopsis Ladanum.
*Linaria minor.
*Papaver Argemone.
*Reseda lutea.
*Scabiosa arvensis.

TOWARDS THE STATION.

*Centaurea Cyanus (roadside).
*Echium vulgare.

Farthing Downs, a mile or so in extent, bare, covered with short turf, and a scanty surface soil over the chalk.

*Cnicus acaulis.
*Plantago media.

HEDGE AND BORDERING BANK LEFT.

*Arabis hirsuta.
*Cistus Helianthemum.
*Daucus Carota.
*Gymnadenia conopsea (also small white variety).
*Linum catharticum.

*Ophrys muscifera.
*Tamus europæus.
*Taxus communis.
*Valeriana officinalis.

BORDERS OF THE WOOD BEYOND.

*Orchis maculata.

Hills on the Coulsdon side; slopes cultivated, wooded above. Oak, Beech, Ash, Black-thorn, Hazel, Viburnum Lantana, Cornus sanguinea, Prunus Cerasus, Clematis.

ON THE BORDERING BANKS.

*Anthyllis Vulneraria.
*Asperula cynanchica.
*Centaurea Scabiosa.
*Daucus Carota.
*Echium vulgare.
*Fragaria vesca.
*Gymnadenia conopsea.
*Hippocrepis comosa.
*Hypericum hirsutum.
*Linum catharticum.
*Onobrychis sativa.
*Orchis pyramidalis.
*Reseda lutea.
*Rubus glandulosus.
*Scabiosa arvensis; *Columbaria.
*Verbena officinalis.

FIELDS NEAR COULSDON.

*Scandix Pecten-Veneris.

Coulsdon Common, a level expanse, overgrown with Furze, Heath, Bracken, and scattered White-thorn. Kenley Common is an open grassy level expanse, with but little Furze upon it. About Coulsdon (in addition to the above),

Aceras anthropophora (?).
Acorus Calamus.
Actinocarpus Damasonium.
Adoxa moschatellina.
Agrostis Spica-Venti.
*Aira flexuosa.
Allium ursinum.
Alsine tenuifolia.
*Anagallis arvensis; cærulea.
Aquilegia vulgaris.
Astragalus Glycyphyllos.
Atropa Belladonna.
*Bunium flexuosum (wood towards Kenley).
Buxus sempervirens.
Carex præcox; *glauca; pilulifera: sylvatica; pendula.
Centunculus minimus.
Cephalanthera grandiflora.
*Chlora perfoliata.
Convallaria majalis.
Corydalis claviculata.
*Cuscuta Epithymum.
Daphne Mezereum.
Dianthus Armeria; deltoides.
Erigeron acris.
*Erythræa Centaurium.
Euonymus europæus.
Euphorbia platyphylla.
Festuca duriuscula.
*Fumaria parviflora.
*Galium tricorne; *cruciatum.
Habenaria viridis.
Helleborus viridis.
Hottonia palustris.
Hypericum Androsæmum.
*Kœleria cristata (Kenley Common).
*Lamium Galeobdolon.
Lathræa squamaria.
*Lathyrus sylvestris.
Leonurus Cardiaca (?).
Limosella aquatica.
Linaria spuria; Elatine.
Luzula Forsteri.
*Melampyrum pratense (wood towards Kenley).
Mentha Pulegium.
Monotropa Hypopitys.
Neottia Nidus-avis.
Ophrys apifera; arachnites (?).
Orchis Morio.
Orobanche minor; elatior.
*Oxalis corniculata.
*Papaver somniferum.
*Petroselinum segetum.
Phyteuma orbiculare.
Pimpinella magna.
*Primula vulgaris; officinalis; elatior.
Pyrus communis.
Radiola Millegrana.
Ranunculus parviflorus; *auricomus.
*Rhamnus catharticus; Frangula.
Ribes rubrum; Grossularia.
*Ruscus aculeatus.
Scilla autumnalis (?).
*Sison Amomum.
Spiræa Filipendula.
Spiranthes autumnalis.

*Thesium linophyllum (towards Kenley).
*Triodia decumbens.
Veronica montana.
*Viola sylvestris ; *hirta ; palustris ; *arvensis ; odorata.

CRYPTOGAMS.

Botrychium Lunaria.
Ophioglossum vulgatum.

Note.—What extent of country the term "about" Coulsdon (D. Cooper) may cover is uncertain; it may include Caterham and Riddlesdown, as well as Farthing Downs and Kenley Common. On Coulsdon Common there are but few moory spots, and only one or two shallow pools. The woods are enclosed.

ADDENDA.

Wood beyond Farthing Downs, near Chaldon, Galanthus nivalis both sides of lane.

29. SUTTON AND BANSTEAD DOWNS, WITH BORDERING FIELDS.

This section of the North Downs extends westwards from Croydon, and attains near Banstead an altitude of five hundred and fifty feet. But the rise is very gradual from the gravelly plain, which intervenes between it and the southern heights of London. Towards Carshalton and Sutton the slopes are cultivated; but the soil is poor and chalky. Banstead Down is furze-grown; in some parts the turf is composed of Helianthemum and much cut up with rabbit holes; westward the furze gives place to Juniper, and beyond this, in the direction of Cheam and Epsom, are cultivated fields. A branch railway from Sutton has been opened up to Banstead and the racecourse on the Epsom Downs. Surface soil, gravel; subsoil, chalky grit with flint.

ROADSIDE FROM CARSHALTON TO BANSTEAD.

*Anthyllis Vulneraria.
Arenaria serpyllifolia (fields).
*Bartsia Odontites (fields).
*Bromus erectus (pit at Carshalton).
*Centaurea Scabiosa; *jacea.
Chelidonium majus (chalk-pit, and about Carshalton, &c.).
*Cichorium Intybus.
*Clematis and *Cornus sanguinea (hedge).
Crepis taraxacifolia (about Sutton).
*Daucus Carota.
Echium vulgare.
Euphorbia exigua.
*Onobrychis sativa.
*Ononis arvensis.
*Origanum vulgare.
*Poterium Sanguisorba.
*Reseda lutea.
*Scabiosa arvensis; *Columbaria (in profusion, pasture left).
Sherardia arvensis (fields).
Silene inflata.
Specularia hybrida (fields).
*Thymus Serpyllum.
Viola tricolor; Valerianella, &c.

ON THE DOWNS, ETC.

*Ajuga Chamæpitys (fields beyond).
*Asperula cynanchica.
*Avena pubescens.
*Briza media.
Calamintha Acinos (fields).
*Carduus acaulis.
*Carex glauca; *flava.
*Carlina vulgaris.
Caucalis daucoides (fields).
Crepis fœtida (?), biennis (?), (olim).
*Erica cinerea.
Euphrasia officinalis.
*Filago germanica.
*Galeopsis Ladanum (fields beyond).
*Gentiana Amarella (scarce).
Habenaria viridis?
*Helianthemum vulgare.
*Linum catharticum.
Onobrychis sativa.
*Polygala vulgaris.
*Poterium Sanguisorba.
Ranunculus parviflorus.
*Rosa micrantha.
*Scabiosa Columbaria.

Scleranthus perennis (fields beyond ?).
*Sedum acre (plenty about Sutton).
*Spiræa Filipendula.
Spiranthes autumnalis.
Thesium humifusum ?
*Viola hirta, var. calcarea ?

Note.—The upper portion of the Downs, where Thesium and the Orchids may have grown, is now cut up by the railway and enclosed.

ADDENDA.

Near Ewell, Sambucus Ebulus; Papaver hybridum.

Epsom Downs: these are a continuation of the Banstead Downs in a south-westerly direction. They are open plains, free from Furze and Juniper. *Carduus acaulis; *Spiræa Filipendula; *Erythræa Centaurium; *Cistus Helianthemum; *Poterium Sanguisorba. In the warren adjoining: *Campanula glomerata; with *Cistus in profusion. Copse towards Ashtead Park: *Campanula Trachelium.

Epsom Common, on the other side of the town, and fronting the chalk-downs, but of lower elevation, is a turfy furze-grown waste; the Furze dense, and mingled with Bracken and Ling; subsoil, gravel. *Salix repens; *Potentilla Tormentilla; *Junci and coarse Grasses, &c.

Fields bordering the Downs at Epsom: Linaria Elatine; Bupleurum rotundifolium ? (*olim*); Myosurus minimus; Fumaria parviflora, &c.

On a farm one and a half miles from Banstead Park, Ashtead Park: Tilia parviflora. Cornfields between Ashtead and Leatherhead: Camelina sativa.

30. DITTON MARSH AND ESHER COMMONS.

Ditton Marsh, about three miles west of Kingston, is traversed at its upper end by the Portsmouth road and South-Western railway. Its longest diameter is from north to south, and, inclusive of Western Green (near the Thames Ditton station, on the branch line to Hampton Court), a mile in length. It is perfectly level, furze-grown for the most part, and generally wet and marshy. A rill runs along its eastern border, and there are ditches and drains by the railway embankment; there are also at this end and on the green adjoining several ponds, but the water on the marsh, at any rate on its lower portion, does not seem to find an outlet in any of these receptacles. Subsoil, gravel.

*Achillea Ptarmica.
*Actinocarpus Damasonium (pond).
*Aira cæspitosa.
*Anthemis nobilis.
Arnoseris pusilla (sandy fields, about).
Barbarea præcox (wastes about Ditton).
*Betonica officinalis.
*Calluna vulgaris.
*Campanula rotundifolia.
*Carduus palustris; *pratensis.
Carex vesicaria (near Esher station); pallescens (copse near Hook).
*Ceratophyllum aquaticum (pond).
*Chenopodium Bonus-Henricus (near Ditton station).
*Erica Tetralix.
*Hieracium umbellatum.
*Hordeum pratense.
*Hydrocotyle vulgaris.
*Inula Pulicaria (roadside); Helenium (meadow near Hook).
*Limosella aquatica (pond between marsh and green).
*Mentha Pulegium.
*Molinia cærulea.
*Myosotis cæspitosa (ditches).
*Myriophyllum alternifolium.
*Œnanthe fistulosa.
*Ononis spinosa.
*Polygonum minus.
*Trifolium subterraneum (Ditton Green).
*Ulex Gallii, etc.

About Thames Ditton.

Alopecurus agrestis (cornfields).
Chenopodium polyspermum (gardens).
Œnanthe Phellandrium (ditches).
Scilla autumnalis.
Sium latifolium (ditch between village and the green).
Solanum nigrum.

Cornfields and Meadows towards Telegraph Hill.

*Agrostis Spica-Venti.
*Anchusa arvensis.
*Antirrhinum Orontium.
*Chrysanthemum segetum.
*Rhinanthus Crista-Galli.
*Sinapis arvensis.

Telegraph Hill, in the wood.

*Circæa lutetiana.
*Fragaria vesca.
Luzula Forsteri.
*Lysimachia nemoralis.
*Valeriana officinalis, &c.

Bordering Fields.

*Bromus secalinus.
*Gastridium lendigerum (scarce).
Hypochœris glabra, &c. (H. C. Watson).
*Silene anglica.
*Spergula arvensis.
*Valerianella dentata.

Lanes and bushy places about Claygate.

*Carex divulsa.
*Euphorbia amygdaloides.
Hypericum Androsæmum.
Iris fœtidissima (?) (sparingly).
Myosurus minimus (fields near Claygate).
*Poa nemoralis.
*Ranunculus auricomus.
Rosa tomentosa.
Ruscus aculeatus.
*Tamus communis.

Esher Common proper, with adjacent Abrook Common and Oxshott Hill, are situated south of Ditton Marsh and to the eastward of Claremont. Collectively they cover a considerable tract of unenclosed heathy land, inclusive, on the Claremont side, of a sandy ridge planted with Fir-trees. The commons are marshy, and drained by the rill above mentioned; there are several swampy pools upon them, of which one has been partially drained. Subsoil, gravel; in some spots sandy or peaty.

It is crossed by a road from Esher to Oxshott, and approached from Ditton by Hare Lane. A road from Esher runs also by the side of the park palings. Oxshott Hill is a sandy heath.

Note.—These plantations of Fir-trees and sandy hills, with occasional intervening sandy fields, extend in a south-westerly direction towards Chertsey and Woking, and are a peculiar feature of this part of Surrey.

*Achillea Ptarmica.
*Aira cæspitosa.
*Alchemilla arvensis (by the park palings).
*Anthemis nobilis.
*Betonica officinalis.
Bunium flexuosum.
*Campanula rotundifolia; Rapunculus (sparingly (?) olim).
*Carduus pratensis; *palustris.
*Carex flava; *vulgaris; *glauca; *panicea; *binervis; *præcox (by the park palings); *pilulifera (sandhill).
*Drosera rotundifolia; *intermedia.
*Epilobium palustre.
*Eriophorum angustifolium.
*Erythræa Centaurium.
*Euphrasia officinalis.
Filago spathulata (about Ditton and Esher).
Gastridium lendigerum (cornfield near Oxshott Hill).
*Genista anglica.
Gentiana Pneumonanthe (sparingly behind farmhouse on the Oxshott road, H. C. Watson).
*Hieracium umbellatum; *vulgatum (roadside hedges towards Oxshott); tridentatum.
*Hydrocotyle vulgaris.
*Hypericum pulchrum; *Elodes; *humifusum; *tetrapterum.
Hypochœris glabra? (by the park palings; olim).
*Juncus squarrosus.
*Leontodon hirtus (by the park palings).
*Linaria Elatine and *minor (cornfield near Oxshott Hill).
Littorella lacustris.
*Lotus major.

*Mentha arvensis (Oxshott Hill; cornfield below).
*Molinia cærulea.
*Myosotis palustris; *cæspitosa.
*Narthecium ossifragum (foot of Oxshott Hill).
*Œnanthe fistulosa.
*Pedicularis sylvestris.
*Pimpinella Saxifraga.
*Potentilla Tormentilla.
*Radiola Millegrana (scarce).
*Ranunculus tripartitus (Hare Lane).
*Rhynchospora alba (foot of Oxshott Hill).
*Salix repens; pentandra (enclosure behind second farmhouse on Oxshott road).
*Scirpus cæspitosus; *setaceus; *fluitans.
*Scutellaria minor; *galericulata.
*Sparganium simplex.
*Tamus communis (hedges).
*Thymus Serpyllum.
*Ulex Gallii.
Vicia lathyroides ?

Beyond Esher and to the right of the Portsmouth road is a hilly slope called Winter Downs, and below is Esher West-end Common, where there are some ponds and some marshy land; on the slope right of the downs proper, is a patch of scrub enclosing a small bog. Drains have been cut across the common, and it is now used as rifle-practising ground; subsoil, gravel; of the hill, sand.

*Actinocarpus Damasonium (lower pond; plenty in 1875).
*Anagallis tenella (bog).
*Carex remota (lane beyond).
*Circæa lutetiana (lane beyond).
*Drosera rotundifolia (bog).
*Erica cinerea; Tetralix.
*Eriophoron angustifolium (bog).
*Hydrocotyle vulgaris.
*Mentha Pulegium (margins of pond).
*Menyanthes trifoliata.
*Myosotis cæspitosa.
*Narthecium ossifragum.
*Ranunculus peltatus; *hederaceus.
*Scirpus acicularis (sandy patches by the lower pond).
*Teesdalia nudicaulis (Winter Downs).
*Valeriana officinalis (lane beyond).

CRYPTOGAMS.

*Aulocomnion palustre (bogs).
*Bartramia pomiformis (bank, further end of Oxshott Hill).
*Bryum erythrocarpum (Winter Downs).
*Chara vulgaris (pond, Weston Green).
*Hypnum stellatum (bog, Winter Downs).
*Leucobryum glaucum (under the Firtrees on ridge, west slope).
Lomaria Spicant.
*Lycopodium inundatum.
*Pilularia globulifera (pond opposite front of Claremont, on Esher Common).
*Pteris aquilina.
*Sphagnum cymbifolium; *acutifolium.

ADDENDA.

Copses beyond Oxshott Hill, near Jacob's Well,* Primula vulgaris in profusion; clover field, Chessington, Crepis fœtida; fourth field from Chessington Church, towards Epsom, Linum angustifolium; near Chessington, and about Hook, Valerianella Auricula.

31. MOULSEY HURST, AND BANKS OF THE MOLE, ABOUT MOULSEY AND ESHER.

Moulsey Hurst is an open grassy plain with a gravelly subsoil, by the Thames, opposite Hampton, and is used on occasion as a race-course. At its eastern extremity are some shallow disused gravel-pits, and near these a marshy patch. The Mole enters the Thames at the bridge, East Moulsey. It is a sluggish winding river, with banks bordered by Willows. The bordering meadows are somewhat damp, but not marshy; subsoil alluvial.

On the Hurst.

*Anthemis nobilis.
*Carduus acaulis.
Carex disticha (H. C. Watson).
Juncus obtusiflorus (marshy corner).
Kœleria cristata (gravel pits).
*Mœnchia erecta.
*Plantago media.
Potentilla argentea.
*Thymus Serpyllum.
*Trifolium subterraneum.
*Scilla autumnalis (near the gravel pits).

Lanes and roadsides between Moulsey and the meadows :—

*Fumaria officinalis.
*Galium cruciatum ; verum.
*Lamium incisum.
*Myosotis collina.

River-side and meadows going by the footpath across the fields to the mills, Esher :—

*Allium vineale.
*Œnanthe crocata (ditch by the path).
*Orchis mascula ; *Morio.
*Poterium Sanguisorba (brought down by floods from the chalk slopes ?).
*Primula veris.
*Ribes nigrum.
*Salix alba ; *fragilis ; *purpurea, var.

Cryptogams.

*Tortula subulata (stumps of trees).

Addenda.

By the Thames, opposite Hampton,* Diplotaxis tenuifolia ; opposite Ditton, Salix rubra ; between E. Moulsey Church and Ember Bridge, Leersia oryzoides ; roadside from bridge to W. Moulsey, Trifolium glomeratum ?

32. BANKS OF THE COLNE, BETWEEN UXBRIDGE AND HAREFIELD.

The Colne flows in a dual stream, with a canal in connexion with it. Its waters are clear, and the current moderately swift. The meadows through which it meanders are generally marshy, and, though drained, are subject to floods when the waters are high. The low hills on the eastern side of the vale about Harefield are wooded ; above, a gravelly subsoil overlies the chalk, but below the chalk is apparent. On the opposite side are low chalk hills at a little distance, free from trees and scrub. Towards Uxbridge this formation disappears. By the canal is a towing-path, and at Harefield are chalk-pits ; Harefield Common, formerly an extensive open district, has all been enclosed ; it possessed a rich and varied flora, of which particulars were recorded by Blackstone (A.D. 1800). Old Park Wood is private property, permission must be obtained to visit it ; Moor Park is free, but there is not much out of the common within its precincts. The lane from Harefield towards Rickmansworth, between the road and the canal, leads to a farm near that town. Bacher Heath, by the high-road, is a small patch of Furze, Heath, Brake, and Bramble, with a moory hollow on its north-west border, and a wood beyond.

Banks of the canal, river, and adjoining moors, ditches, &c. :—

Acorus Calamus.
Actinocarpus Damasonium ? (ponds about Uxbridge), olim.
Alisma ranunculoides.
*Alyssum incanum (two or three plants ; may be more ; casual).
*Angelica sylvestris.
*Arctium majus.
*Artemisia vulgaris.

*Butomus umbellatus.
*Caltha palustris.
*Cardamine amara (Harefield)
*Carex riparia; *paniculata; *vulpina.
*Digraphis arundinacea.
Drosera rotundifolia (bogs, Uxbridge).
*Epilobium hirsutum; *parviflorum.
*Eupatorium cannabinum.
*Galeopsis Tetrahit.
*Glyceria aquatica.
*Helosciadium nodosum.
*Hippuris vulgaris.
*Hypericum tetrapterum.
*Hypochœris Morsus-ranæ.
*Impatiens fulva.
Inula Helenium (meadows near Harefield, and at Breakspears).
*Iris Pseudacorus.
*Lysimachia vulgaris; Nummularia.
*Lythrum Salicaria.
*Malva moschata.
*Molinia cærulea.
Narcissus Pseudo-Narcissus.
*Nuphar lutea.
Nymphæa alba.
*Œnanthe Phellandrium; fluviatilis; *fistulosa.
Orchis latifolia.
Ornithogalum umbellatum ?
Parnassia palustris (Harefield).
*Pedicularis sylvatica; palustris.
Petasites vulgaris.
*Phragmites vulgaris.
*Polygonum amphibium.
Ranunculus Lingua (Uxbridge Moor).
*Rumex Hydrolapathum.
*Sagina nodosa (canal-side and moor).
*Sagittaria sagittifolia.
*Salix alba; *fragilis; *cinerea; *viminalis; pentandra (near the brick-kiln).
Sambucus Ebulus (Uxbridge Moor).
*Scutellaria galericulata.
*Senecio aquaticus.
*Silaus pratensis (abundant).
*Sium angustifolium (abundant).
*Solanum Dulcamara.
*Sparganium ramosum.
*Spiræa Ulmaria.
*Stachys palustris.
*Stellaria aquatica (abundant).
*Tanacetum vulgare.
*Thalictrum flavum (scarce).
*Triglochin palustre.
*Viburnum Opulus.

Woods, banks, roadsides, chalk-pits, &c.

Aceras anthropophora ?
*Achillea Ptarmica (Bacher Heath).
*Adoxa moschatellina.
Alopecurus agrestis (fields).
*Asperula odorata.
Astragalus Glycyphyllos.
*Betonica officinalis.
Buxus sempervirens.
Calamagrostis Epigejos.
*Calamintha Clinopodium; *Nepeta (lane leading to the river, a plant or two); *Acinos.
*Campanula glomerata.
*Carduus acaulis; nutans.
*Carex sylvatica.
Carlina vulgaris.
*Chlora perfoliata.
*Clematis Vitalba.
*Daphne Laureola.
Dentaria bulbifera (Old Park Wood plenty, and Garrett Wood).
*Digitalis purpurea.
*Erythræa Centaurium.
*Euonymus europæus.
*Euphorbia amygdaloides; exigua.
*Euphrasia officinalis (Bacher Heath).
Galium cruciatum.
*Genista anglica (Bacher Heath).
*Gentiana Amarella (plentiful).
*Geranium dissectum; lucidum (lane leading to the river).
Gnaphalium sylvaticum (groves, Bartleswell).
Gymnadenia conopsea.
Hyoscyamus niger.
*Hypericum perforatum; *hirsutum; Androsæmum.
*Inula Conyza.
*Lamium Galeobdolon.
Lathræa Squamaria (lane leading to the river).
Limosella aquatica (Warren pond, Breakspears).
*Linaria vulgaris; *spuria; *Elatine; *minor.
*Linum catharticum.
Listera ovata.
*Luzula pilosa.
*Lysimachia nemorum.
*Melilotus officinalis; *alba.
*Mentha Pulegium; sativa (Bacher Heath).
*Molinia cærulea (Bacher Heath).
Neottia Nidus-avis (Whiteheath Wood).
*Ononis arvensis.
*Orchis mascula; *pyramidalis; militaris (?); ustulata.
*Origanum vulgare.
*Oxalis Acetosella.
Paris quadrifolia (Old Park Wood).
*Picris hieracioides.
*Poterium Sanguisorba.
*Primula veris (fields); *vulgaris.
Pyrus Aria; Aucuparia.
*Reseda lutea; *Luteola.
Rhamnus catharticus (hedges); Frangula.
Ribes rubrum and nigrum (meadows and copses Warren pond, Breakspears).

WHITE HEATH WOOD.

Sambucus Ebulus (meadow at Break-spears).
*Sanicula europæa.
*Scilla nutans.
Thymus Serpyllum.
*Ulex Gallii (Bacher Heath).
*Viburnum Lantana.
Viola hirta.

CRYPTOGAMS.

Aspidium aculeatum (Old Park Wood).
Equisetum sylvaticum.
Eucalypta vulgaris.
Nephrodium cristatum (Bartleswell).

33. COLNEY HEATH.

North of the ridge referred to in preceding sections as extending eastwards and westwards of Stanmore Heath, there are no indications of more than a very gradual inclination towards Hatfield and St. Albans. About midway between these towns, and a mile or so from the Spring Field station on the branch line of railway which connects them, is an unenclosed patch of common, partly furze-grown, known as Colney Heath; the sluggish rivulet which drains it is a feeder of the Colne. The higher furze-grown portion is on the western border; the banks of the rivulet are swampy; subsoil, gravel. Shady lanes lead to it from different directions, and in the immediate neighbourhood are N. Mimms and Tittenhanger Parks; adjoining the common, and opposite a windmill, is a detached heathy portion, upon which are some pools.

*Bidens tripartita.
*Calluna vulgaris.
Carduus pratensis; acaulis.
Carex pilulifera; vesicaria ?
Carlina vulgaris ?
Centunculus minimus ?
*Epilobium palustre; obscurum.
*Euphrasia officinalis.
*Gentiana campestris (among the Furze, end towards the windmill).
*Helosciadium nodiflorum; *inundatum.
*Hydrocotyle vulgaris.
*Inula Pulicaria.
Lythrum hyssopifolium ? olim (and in an adjoining field).
*Mentha Pulegium.
*Menyanthes trifoliata.
Mœnchia erecta.
*Œnanthe fistulosa.
*Polygonum minus.
Potamogeton rufescens; acutifolius.
Radiola Millegrana.
Scutellaria minor.
*Sparganium simplex.
Triglochin palustre.
Ulex Gallii.

Neighbouring lanes, &c., towards Radlett, N. Mimms, Springfield, St. Albans.—

*Betonica officinalis.
*Calamintha Clinopodium.
*Campanula rotundifolia.
*Circæa lutetiana.
*Clematis Vitalba.
*Digitalis purpurea.
*Euphorbia exigua (cornfields).
*Galeopsis Tetrahit.
*Lysimachia Nummularia.
*Malva moschata.
*Mentha sativa.
*Ononis arvensis.
*Rhamnus catharticus.
*Scabiosa arvensis.
*Tamus europæus.
*Verbascum Thapsus (lanes towards Springfield and Hatfield, plenty).

CRYPTOGAMS.

Aspidium angulare (towards N. Mimms).
*Chara flexilis (on the heath in a pool).
Fissidens taxifolium (Colney Street).
*Nephrodium Filix-mas.

ADDENDA.

*Setaria viridis (Springfield station).
*Chrysanthemum segetum (cornfield near).
Crocus vernus (Brookman's Park, N. Mimms).

34. BROXBOURNE AND WORMLEY WOODS.

Northaw Common, near the Potter's Bar station on the Great Northern line, is now enclosed; these woods are therefore better and more directly approached from Broxbourne, on the Great Eastern Railway. They cover the low hills which border the Lea Valley on the west, and are only partially enclosed. They are of a cold damp character, with a gravelly subsoil, but with much alluvial deposit in the hollows, and along the lines of drainage to the river Lea. Roads, leading from Broxbourne to Bayford and Brickendon Green, intersect them from east to west; the more direct approach is through the grounds of Broxbourne Park; the longer route is *viâ* Wormley, west end. Or that from Cheshunt, *viâ* Applebury and Beaumont Green, may be selected. Oak is the prevailing timber, but there is no lack of scrub and of other kinds of trees, Hornbeam, Beech, Ash, &c. There are indications of underlying chalk.

*Arctium majus.
*Bartsia Odontites (lanes).
*Betonica officinalis.
*Carex pendula (abundant); pallescens; strigosa.
*Chlora perfoliata (in a boggy field, left of the road to Bayford); a few plants.
*Chrysanthemum segetum (fields near Wormley west-end, in profusion).
Chrysosplenium oppositifolium.
Cicuta virosa (in a swamp near the western border of Wormley Wood; also in a swampy pool near Applebury; and formerly in ponds near Brickendon Green).
*Circæa lutetiana.
Convallaria majalis.
*Digitalis purpurea.
*Epilobium angustifolium (also in Callis wood); *montanum.
*Epipactis palustris (boggy field above mentioned).
*Eriophoron angustifolium (boggy field above mentioned).
*Erythræa Centaurium (abundant).
*Euphrasia officinalis (luxuriant).
*Genista tinctoria (meadows left of the road to Bayford, also on the western borders of the wood).
*Gymnadenia conopsea (in the boggy field above mentioned—a few plants).
Habenaria viridis (meadow near W. border of Wormley Wood).
*Hypericum Androsæmum; *perforatum; *tetrapterum; *hirsutum; *humifusum.
*Lysimachia nemorum; *Nummularia.
*Mentha sativa.
*Menyanthes trifoliata (bog with Cicuta).
*Orchis maculata (meadows in the gorge near Wormley west-end).
*Rhinanthus Crista-Galli (meadows in the gorge near Wormley west-end).
*Ruscus aculeatus.
*Triglochin palustre (boggy meadow).
Utricularia vulgaris (pond).
*Valeriana officinalis.
*Verbascum Thapsus (near Beaumont Green).

CRYPTOGAMS.

*Chara hispida (boggy meadow).
*Equisetum maximum (abundant).
*Nephrodium Filix-mas.

ADDENDA.

IN THE RIVER AT BROXBOURNE, AND BORDERING DITCHES, &c.

*Carex Pseudo-cyperus; strigosa (woods).
*Œnanthe fluviatilis.
*Salix viminalis; *cinerea; *fragilis, &c.
*Potamogeton lucens; *densus; *perfoliatus.
*Typha latifolia.

Nasing Wood, east of Broxbourne: Lathyrus hirsutus. Cornfields, Hoddesdon: Delphinium Ajacis.

Marshes at Hoddesdon: Carex pulicaris; acuta; disticha; ampullacea; Utricularia vulgaris; Fritillaria Meleagris (?); Zannichellia palustris. Ditches: Myriophyllum verticillatum; Anagallis tenella; Hottonia palustris; Stellaria glauca; Hippuris vulgaris; Mentha sylvestris (bank of the stream near railway).

Alyssum calycinum (banks of the New River).

Myosurus minimus (between Hoddesdon and Rye House); also Allium vineale and Avena strigosa.

About Stanstead: Salix Lambertiana. Between Stanstead and Ware, by the towing path: Camelina sativa. Roydon Lane: Dipsacus pilosus. Near Stanstead: Hyoscyamus niger. Between Stanstead and Rye House: Hippuris vulgaris.

Hoddesdon, west field, road to Hertfield: Orobanche cærulea on Achillea Millefolium.

35. WARLEY COMMON.

Warley Common near Brentwood, like many other localities about London, has undergone alteration of late years; much of the bordering forest-land has been enclosed, and a gravelly pasture fronting the depot barracks has been converted into a drilling ground for the recruits. The common is about one hundred acres in extent, its longest diameter from north to south. The woods, which surround it on all sides but the south, are mostly of Oak, Birch, and Hornbeam, with an undergrowth of Bracken, Hazel, Thorn, &c. In the centre, and opposite the entrance to the grounds of Thorndon Hall, is a copse, and in this direction the common is covered with a dense scrub of Oak and Hornbeam. In more open parts, Bracken and Ling prevail. The woods are damp, and in parts with an undergrowth of Fern, occasionally swampy, with Elders and Grey Sallows on their borders. A ravine separates the upper part of the common from a grass-plot sloping to the south. It receives the drainage of the swampy woods on the western border, and its sides are clothed with Briars, Brambles, Sallows, Furze, and Thorn; subsoil, gravel.

ON THE COMMON OR IN THE SURROUNDING WOODS AND NEIGHBOURHOOD.

Achillea Ptarmica.
Adonis autumnalis (fields near).
*Aira flexuosa; *cæspitosa.
Alisma ranunculoides.
Anagallis tenella.
Anthemis nobilis.
Arctium majus (woods).
Betonica officinalis.
Bunium flexuosum.
Campanula Trachelium.
Carex divulsa; muricata;* panicea; stellulata; ovalis; vulgaris; pendula; Œderi; binervis; lævigata; hirta; Pseudo-cyperus.
Carlina vulgaris (?).
Cuscuta Epithymum.
*Digitalis purpurea.
Epipactis latifolia (?).
*Erica Tetralix; *cinerea.
Eriophoron angustifolium.
Euphrasia officinalis.
Filago minima.
Galeopsis Tetrahit.
Galium saxatile; uliginosum; tricorne (?) (cornfields near).
Gnaphalium sylvaticum.
Hieracium umbellatum; *vulgatum.
*Hydrocotyle vulgaris.
*Hypericum hirsutum; Elodes; pulchrum.
Ilex aquifolia.
Iris fœtidissima.
*Juncus supinus; lamprocarpus; *squarrosus; *obtusiflorus.
Lactuca virosa.
Lathyrus Nissolia.
Lemna polyrhiza.
Leontodon hirtus.
Listera ovata.
Luzula sylvatica.

*Melampyrum pratense.
*Mentha sativa; *hirsuta.
Menyanthes trifoliata.
Mœnchia erecta.
*Molinia cærulea.
*Nardus stricta.
Orchis maculata.
Polygonum minus.
Populus tremula.
*Pyrus Malus.
Rhamnus Frangula.
*Rosa canina.
Rubus (several segregates; see Index).
Ruscus aculeatus.
Taxus baccatus.
*Teucrium Scorodonia.
Thymus Serpyllum.
Trifolium (subterraneum); ochroleucum (in a fish pond, Heron).
*Triodia decumbens.
*Salix repens; *Caprea; *cinerea.
Samolus Valerandi.
*Sarothamnus scoparius.
*Scabiosa succisa.
Scirpus cæspitosus; setaceus.
Scutellaria galericulata.
Senecio erucifolia.
*Solidago Virgaurea.
Spiranthes autumnalis.
Stachys Betonica; palustris.
Stellaria uliginosa.
Valeriana dioica.
Vicia hirsuta.
Vinca minor (thicket, S. weald, Little Warley).
Viola palustris (boggy thicket).

CRYPTOGAMS.

Aspidium aculeatum and var. lobatum.
Athyrium Filix-fœmina.
Equisetum sylvaticum.
Hypnum fluitans, &c.
Lomaria Spicant.
Nephrodium dilatatum; spinulosum; Oreopteris.
Ophioglossum vulgare (Great Warley).
Osmunda regalis (grew here formerly in the wood west of the common).
Polytricha.
Sphagnum cuspidatum.

36. PURFLEET.

The flats by the Thames, which extend thus far from the eastern districts of London on the Essex shore, are here broken for a short space, by the intervention of a low chalk ridge, which extends for some distance beyond, in the direction of Grays and Tilbury; with flats again, however, between it and the river. Here are old and extensive excavations, and a low hill with overlying gravel drift, well wooded. Many plants of the chalk formation occur, together with others peculiar to the muddy shores of a salt-water creek.

Ajuga Chamæpitys (cornfields near).
Anthemis arvensis (cornfields near).
Armeria maritima.
Artemisia Absinthium (?).
*Aster Tripolium.
*Atriplex littoralis; *portulacoides.
*Bupleurum tenuissimum; rotundifolium (?) (cornfields).
*Calamintha Clinopodium; Acinos (cornfields).
Carduus nutans.
Centaurea Calcitrapa.
Chenopodium olidum; ficifolium.
Cichorium Intybus.
*Clematis Vitalba.
Crepis taraxacifolia; fœtida; biennis.
Cynoglossum officinale; montanum.
*Daucus Carota.
*Diplotaxis tenuifolia.
Erigeron acris.
*Erythræa Centaurium; pulchella.
*Euphrasia officinalis.
*Festuca Myurus.
Fœniculum vulgare.
Galeopsis Ladanum (cornfields near).
Galium parisiense (?) anglicum.
Geranium columbinum.
*Hordeum maritimum.
Hydrocharis Morsus-ranæ (ditches near).
Inula Conyza.
Iris fœtidissima (wood).
Juncus maritimus.
Lactuca saligna.
Leontodon hirtus.
Lepidium campestre.
*Lepturus filiformis.
Linum angustifolium; catharticum.
Lithospermum officinale.
Marrubium vulgare.
*Medicago maculata.
*Melilotus officinalis.
*Mentha rubra (ditches near).
Nepeta Cataria.
*Œnanthe Lachenalii.
Onopordum Acanthium.

Orchis pyramidalis.
*Origanum vulgare.
*Petroselinum segetum.
*Phragmites communis (ditches).
*Picris hieracioides.
Reseda lutea.
Rumex maritimus; palustris; *Hydrolapathum.
Salicornia herbacea.
*Salvia verbenaca.
*Saponaria officinalis (wood).
Saxifraga granulata (Woodford wood).
Scabiosa Columbaria.
*Sclerochloa maritima; distans.
*Senecio erucifolius.
*Sinapis nigra.
Smyrnium Olusatrum.
Solidago Virgaurea (woods).
Specularia hybrida (cornfields).
*Statice Armeria.
*Suæda maritima.
Tilia parvifolia (Woodford wood).
*Triglochin maritimum; palustre.
Verbascum Thapsus; nigrum.
*Verbena officinalis.
*Viburnum Lantana.
Vicia hirsuta; *Cracca.

Note.—Polypogon monspeliensis (formerly, on the other side of the ditch towards Raynham); also Tragopogon porrifolius, &c.; Ophrys apifera and Aceras; chalk-pits.

37. GREENHITHE AND DARTFORD.

Opposite Purfleet, and on the other side of the river, are the Stone and Dartford Marshes, similar in character to those between Erith and Woolwich, backed by the chalk slopes about Dartford and Greenhithe, and traversed by the joint current of the Darne and Cray which enters the Thames directly opposite Purfleet. Dartford lies at some little distance back from the banks of the river, but this makes here a reach to the south-east, and comes close up to the chalk banks at Greenhithe further down. The vegetation is similar in character to that of Purfleet; subsoil of the flats, alluvial.

The chalk-pits about Greenhithe are of vast extent, and but for their surroundings the older ones have little in their aspect to indicate their origin. Fields, orchards, habitations, occupy their areas; and one of them is crossed by the North Kent line of railway.

Marshes and Riverside.

*Apium graveolens.
*Artemisia maritima.
*Aster Tripolium.
*Atriplex littoralis; portulacoides.
*Beta maritima.
*Carex divisa.
*Ceratophyllum aquaticum.
*Cynoglossum officinale (ditch banks).
*Festuca Myurus.
*Glaux maritima.
*Hordeum maritimum.
*Hydrocharis Morsus-ranæ.
*Myriophyllum spicatum.
*Plantago maritima.
*Samolus Valerandi.
*Scirpus maritimus.
*Sclerochloa maritima.
*Spergularia neglecta.
*Statice Limonum.
*Suæda maritima.
Trifolium maritimum.
*Triglochin maritimum; palustre.
Triticum junceum.

Roadsides, Fields, Chalk-pits, and Banks.

The country about Dartford is elevated to a small extent only above the level of the flats, with a gravelly subsoil overlying chalk; chalky in many places.

Aceras anthropophora (Greenhithe pits).
Adonis autumnalis (cornfields ?).
*Agrimonia Eupatoria.
Ajuga Chamæpitys (fields).
*Anchusa arvensis.
*Anthyllis Vulneraria.

*Artemisia vulgaris; Absinthium (pits).
*Arenaria trinervis; serpyllifolia.
*Asperula cynanchica.
Avena pubescens.
Bartsia Odontites.
*Briza media.
*Bromus secalinus; *arvensis.
*Bryonia dioica.
Bunium flexuosum (wood, near).
Bupleurum rotundifolium (?).
*Calamintha Acinos; *Clinopodium.
Campanula glomerata.
*Carduus nutans; *acaulis; eriophorus (Stone pit ?).
*Caucalis daucoides (fallow field near Dartford).
*Centranthus ruber (pits—plentiful).
*Chlora perfoliata (pits).
Chrysanthemum Leucanthemum.
*Cichorium Intybus.
*Cistus Helianthemum.
*Clematis Vitalba.
Conium maculatum (?).
*Cornus sanguinea.
*Crepis taraxacifolia; biennis; fœtida.
Cynoglossum officinale.
*Daucus Carota.
Dianthus Armeria.
Digitalis purpurea (Stone chalk-pit).
*Diplotaxis tenuifolia.
*Dipsacus sylvestris.
*Echium vulgare.
Erigeron acris.
Euonymus europæus (Stone pit).
Eupatorium cannabinum (Stone pit).
*Euphrasia officinalis.
*Filago germanica.
*Fumaria parviflora.
Galeopsis Tetrahit.
*Galium verum; *cruciatum; *Mollugo; anglicum (old walls at Dartford (?); olim).
Gentiana Amarella.
*Hippocrepis comosa (road from Dartford to Greenstreet Green).
*Humulus Lupulus.
*Hypericum perforatum.
*Inula Conyza.
Juniperus communis (road from Dartford Heath to Greenstreet Green).
*Lactuca virosa (gravel pits near Dartford).
*Ligustrum vulgare.
*Linaria vulgaris; minor (pits).
*Linum catharticum.
Listera ovata (Stone chalk-pit).
*Lithospermum arvense (near Greenhithe).
*Lotus corniculatus; *tenuis (road to Greenstreet Green from Dartford).
Mentha sylvestris and rotundifolia (between Crayford and Dartford); olim, Pulegium.
Nepeta Cataria.
*Onobrychis sativa.
*Ononis arvensis.
Ophrys apifera; muscifera (Stone chalk-pits (?) olim).
*Orchis pyramidalis (Greenhithe pits); maculata (Stone pit).
*Origanum vulgare.
*Papaver Argemone; *hybridum; *dubium; *Rhœas; *somniferum.
*Parietaria diffusa (Greenhithe).
*Pastinaca sativa.
*Picris hieracioides.
*Pimpinella Saxifraga; magna (?) (park at Greenhithe ?).
Polygala vulgaris.
*Potentilla anserina (gravel pit).
*Poterium Sanguisorba.
Primula vulgaris (Stone chalk-pit).
*Reseda lutea; *Luteola.
*Salvia verbenaca.
*Saxifraga granulata (roadside to Greenstreet Green).
*Scabiosa arvensis; *Columbaria.
*Sherardia arvensis.
*Silene inflata.
*Smyrnium Olusatrum (pits, scarce).
*Tamus communis.
*Thymus Serpyllum.
Trifolium scabrum; subterraneum; striatum.
*Verbascum Lychnitis (frequent); *nigrum; *Thapsus.
*Viburnum Lantana.
*Vicia Tetraspermum; lathyroides (banks about Greenhithe).
Viola hirta (pits).

DARTFORD HEATH.

Common heath plants, Furze, Ling, Bracken, Polytricha, with scattered thorns. On the borders, *Onopordum Acanthium, frequent. Between Dartford Heath and Greenstreet Green, Galeopsis ochroleuca (?) *olim.* Spiranthes autumnalis; *Trifolium striatum; Cynoglossum officinale; chalk-pit near (?). Bexley Heath exists no longer but in name. In the Strawberry beds on the slopes, Arnoseris pusilla and Senecio viscosus were formerly in plenty. Borders of field below: *Setaria viridis;* Mentha arvensis;

*Centaurea Cyanus. Road from Dartford Heath to Greenstreet Green.
Sanguisorba officinalis (?) *olim.*

38. DARNE OR DARENT WOOD.

This wood, three miles south-east of Dartford, is of considerable extent, and covers a hilly ridge extending in an easterly direction for about two miles towards Greenhithe. The subsoil is gravel above, and chalky on the flanks. Remains of an ancient entrenchment are perceptible in places along the crest of the hill. Oak and Beech are the prevailing timber, but there is no lack of other descriptions.

*Agrimonia Eupatoria.
*Ajuga reptans.
*Anemone nemorosa.
*Asperula odorata.
*Betonica officinalis.
Campanula glomerata; *rotundifolia.
*Circæa lutetiana.
*Cistus Helianthemum.
*Clematis Vitalba.
*Convallaria majalis.
*Cornus sanguinea.
Dianthus Armeria.
*Digitalis purpurea.
*Echium vulgare (borders).
*Epilobium montanum.
*Erythræa Centaurium.
*Euphorbia amygdaloides.
*Fragaria vesca (in great profusion).
*Galium cruciatum.
*Geranium columbinum.
*Gnaphalium sylvaticum.
*Habenaria bifolia.
Hypericum Androsæmum; *pulchrum; tetrapterum; *humifusum.
*Iris fœtidissima.
*Lamium Galeobdolon.
*Listera ovata.
*Lithospermum officinale (borders towards Greenhithe); purpureo-cæruleum (towards Greenhithe); arvense (bordering fields).
*Luzula sylvatica.
*Lysimachia nemorum; *Nummularia.
*Melampyrum pratense.
Melica uniflora.
*Mercurialis perennis.
Myosotis sylvatica; *versicolor; palustris.
*Œnothera biennis (borders).
*Orchis mascula; *maculata.
*Origanum vulgare.
*Orobus tuberosus.
*Papaver somniferum (bordering cornfield towards Greenhithe, plenty).
*Primula vulgaris.
Rhamnus catharticus.
*Rosa systyla.
*Ruscus aculeatus.
*Sanicula europæa.
Scrophularia nodosa.
Silene italica.
Solidago Virgaurea.
*Tamus communis.
Veronica officinalis.
*Viburnum Opulus; Lantana.
*Vicia sepium; hirsuta; tetrasperma.
Viola sylvestris.

OLD CHALK-PIT TOWARDS GREENHITHE.

*Astragalus Glycyphyllos.
Galeopsis ochroleuca (?) olim.

CRYPTOGAMS.

Asplenium Adiantum-nigrum.
Athyrium Filix-fœmina.
*Lomaria Spicant.
*Nephrodium Filix-mas.
*Ophioglossum vulgatum.
*Pteris aquilina.
*Woodland mosses in abundance and variety: Hypnum molluscum; curvatum; loreum, &c.

Note.—Orchis hircina was formerly found on the borders of this wood.

39. NORTH DOWNS, NEAR SEVENOAKS.

Opposite Sevenoaks, the North Downs recede from their general direction east and west and open inwards, and in the gap thus formed, the Darent, rising near Westerham, and receiving the drainage of this section of the hills, pursues its course to the Thames. On one side of the funnel in front of Sevenoaks, East Hill projects as a prominent bluff, and on the other, Morant's Court Hill with the famous Knockholt Beeches in its proximity. The general character of these chalk downs, here as elsewhere along the range, is an average altitude of four or five hundred feet above the level of the Weald; facing them, at various distances, is a range of red sandhills; above and beyond the slopes is an open undulating country with cultivated fields, and plantations.

Alchemilla vulgaris (Hill Park, near Westerham).
Anthemis arvensis.
*Anthyllis Vulneraria.
*Asperula cynanchica; *odorata (woods).
Astragalus Glycyphyllos.
Atropa Belladonna (Kemsing, near Wrotham).
*Avena pubescens.
*Brachypodium pinnatum (downs).
*Briza media (downs).
*Bromus erectus (downs).
*Calamintha Clinopodium; *menthifolia; *Acinos.
Campanula glomerata (Morant's Court Hill).
*Carduus acaulis.
*Carlina vulgaris.
*Centaurea Scabiosa; *Jacea.
*Cephalanthera grandiflora (woods).
*Chlora perfoliata.
*Cichorium Intybus.
*Cistus Helianthemum.
*Clematis Vitalba.
Crepis biennis (Morant's Court Hill); fœtida; taraxacifolia.
*Daphne Laureola (woods).
*Daucus Carota.
*Digitalis purpurea (woods).
*Echium vulgare.
*Epipactis latifolia (beechwoods; plenty about Knockholt Beeches).
*Erigeron acris (Shoreham, &c.).
*Erythræa Centaurium.
*Euphorbia amygdaloides.
*Euphrasia officinalis.
*Galeopsis Ladanum (chalky fields).
*Gentiana Amarella (abundant).
*Helminthia echioides (gravel soil beyond the chalk at St. Clare).
*Heracleum Sphondylium (old pit on Morant's Court Hill).
Herminium Monorchis (old pit on Morant's Court Hill).
*Hippocrepis comosa (banks).
Hordeum sylvaticum (thicket by roadside, River Hill, Sevenoaks).
*Hypericum hirsutum; montanum (near Bussels Green, Sevenoaks).
*Inula Conyza.
*Juniperus communis.
*Lactuca muralis (beechwood at Shoreham, abundant).
*Leontodon hispidus.
*Linum catharticum.
*Listera ovata (woods).
*Lotus corniculatus.
*Melampyrum pratense (woods).
*Melica uniflora (woods).
*Mercurialis perennis (woods).
*Ononis arvensis.
*Ophrys apifera (downs, Shoreham, also Morant's Court Hill); aranifera; muscifera (downs near Sevenoaks).
*Orchis pyramidalis (downs); *mascula (woods).
*Origanum vulgare.
*Pastinaca sativa.
Phyteuma orbicularis (Morant's Court Hill).
*Picris hieracioides.
Pimpinella magna (about Sevenoaks and Westerham).
*Plantago media.
*Poterium Sanguisorba.
*Primula vulgaris (woods).
*Reseda lutea; *Luteola.
*Rosa micrantha.
*Rubus corylifolius; glandulosus.
*Scabiosa arvensis; *Columbaria.
*Scilla nutans (woods).
Senecio erucifolius; sylvaticus.
*Sherardia arvensis (fields).
*Specularia hybrida (fields).
Spiræa Filipendula (Morant's Court Hill and downs about Wrotham).
*Taxus communis.
*Thymus Serpyllum.
Verbascum Lychnitis (Morant's Court Hill).
*Viburnum Lantana.
*Viola hirta; *sylvestris.

CRYPTOGAMS.

Anomodon viticulosus (Morant's Court Hill).
Hypnum molluscum; abietinum (Morant's Court Hill).
Neckera crispa (Morant's Court Hill).
*Nephrodium Filix-mas (woods).

ADDENDA.

About Farningham: Anagallis cærulea; Cynoglossum officinalis; Erigeron acris; Verbascum nigrum and Lychnitis; Nepeta Cataria. Between

Down and Cudham: Ajuga Chamæpitys; Habenaria bifolia; Ophrys apifera; Orchis mascula; Lithospermum officinale. Downs at Brasted, Gentiana Amarella.

40. REIGATE HILL AND THE WRAY COMMON.

Conspicuous as a projecting headland, on the chalk range is Reigate Hill, and the adjacent Wray Common. They are in immediate proximity to the town of Reigate, and the extensive excavations on their flank are discernible for a considerable distance. They are separated from each other by a plantation, and adjoin the grounds of Gatton Park, in the direction of Merstham. The plateaux above are crowned with beech-woods, beyond these are cornfields, and cornfields are also general along the base of the slope westwards; subsoil, chalk; below chalk grit, above chalk and gravel. The characteristic plants in abundance.

*Aceras anthropophora (frequent).
*Ajuga Chamæpitys (cornfields).
*Anthyllis Vulneraria.
Arabis hirsuta.
*Asperula cynanchica.
*Avena pubescens.
*Briza media.
*Bromus erectus.
*Calamintha Clinopodium.
*Campanula Trachelium (woods).
*Carduus acaulis; nutans; tenuiflorus.
*Cephalanthera grandiflora (woods).
*Chlora perfoliata.
*Cichorium Intybus.
*Cistus Helianthemum.
*Clematis Vitalba.
Colchicum autumnale (meadow adjoining Wray Common).
*Cornus sanguinea.
*Daucus Carota.
*Echium vulgare.
*Epipactis latifolia (beechwoods).
*Erigeron acris.
*Erythræa Centaurium.
*Euphorbia platyphylla (cornfields).
*Euphrasia officinalis.
*Galeopsis Tetrahit; *Ladanum (cornfields abundant).
*Galium tricorne (cornfields).
*Gentiana Amarella.
Geranium columbinum; pratense (foot of the hill).
Herminium Monorchis (above the pit on the common).
*Hippocrepis comosa (in profusion).
*Inula Conyza.
*Juniperus communis.
*Kœleria cristata.
*Linaria minor; *Elatine and *spuria (in cornfields).
*Linum catharticum.
*Lotus corniculatus.
*Mentha arvensis (cornfields).
*Monotropa Hypopitys (beechwood).
*Neottia Nidus-avis (beechwood).
*Onobrychis sativa.
*Ononis arvensis.
*Ophrys apifera (banks upper end of the road above the Wray pit); muscifera? (banks below the Reigate pits, near the reservoir).
*Orchis pyramidalis; *conopsea; mascula (copses, Wray Common).
*Origanum vulgare.
*Pimpinella magna (wood W. end of the hill).
*Poterium Sanguisorba.
*Prunus Cerasus.
*Reseda lutea; Luteola.
*Rosa micrantha.
*Senecio erucifolius.
*Specularia hybrida (cornfields).
*Spiranthes autumnalis (E. end of Reigate Hill).
*Stachys arvensis (cornfields below).
Thlaspi arvense (cornfields below).
*Thymus Serpyllum.
Trifolium subterraneum (Redhill).
*Viburnum Lantana.

ADDENDA.

In Gatton Park woods, Paris quadrifolia; by the engine pond, Sambucus Ebulus; chalk hill west of Reigate Hill,* Orchis ustulata; lanes below slopes west of Reigate Hill,* Pimpinella magna; roadside between Upper Gatton and Chipstead, Myrrhis odorata; woods about Chipstead, Cephalanthera grandiflora; Epipactis latifolia; Helleborus viridis (wood near Chipstead Church); Lathræa Squamaria; wet copses near, Cardamine amara.

41. REIGATE HEATH, REDHILL AND EARLSWOOD COMMONS.

Fronting the section of the chalk range referred to in the foregoing locality are the hills of the Weald; but this range forms by no means a continuous line similar to that of the downs. Although they bear occasionally the character of a ridge, they are generally grouped in an irregular manner. Here for instance, at Reigate Heath there is a break for some distance in the direction of Dorking. The heath lies left of the road from Reigate to the last-mentioned town. One of the two low hills upon it has a windmill on its summit; the other is crowned with a clump of Fir-trees. At one end of the heath is a bog adjoining a marshy meadow, with a swampy copse near its south-eastern corner; at the other end is another bog, also with an adjoining swampy wood; furze- and heath-grown with a sandy subsoil. Redhill Common, south of the town, is a hill of red sand; furze- and fern-grown, but not heathy. Earlswood Common is continuous with it, but low-lying. It is a grassy and marshy flat, with some ditches and a reservoir at the lower end; subsoil, sand and gravel.

*Alnus glutinosa (swampy copses).
*Anagallis tenella (bogs).
Bupleurum falcatum (sparingly W. end of the Heath (?), olim).
Carex pulicaris; curta; pilulifera; *ampullacea; *paniculata (swampy copse near S.E. corner); hirta (Earlswood Common).
Centunculus minimus.
*Chrysanthemum segetum (cornfield).
*Chrysosplenium oppositifolium (swampy copse near S.E. corner, with *alternifolium, scarce, and copse near Wonham).
Comarum palustre (?) (swampy copses).
*Corydalis claviculata (swampy copses).
*Cuscuta Epithymum; Trifolii.
Delphinium Ajacis (cornfields about).
Dianthus Armeria (lane by Trumpet's Hill).
*Drosera rotundifolia.
Epilobium angustifolium (W. end of Reigate Park).
*Erigeron acris.
*Eriophoron vaginatum; *angustifolium.
Festuca sciuroides (Redhill).
*Filago minima.
*Hydrocotyle vulgaris.
Hyoscyamus niger (also on Redhill).
Hypericum Androsæmum and calycinum (W. end of Reigate Park); *Elodes (bog).
Iberis amara (E. end of Heath ?).
Inula Pulicaria (Earlswood Common).
*Leontodon hispida.
Lepidium Smithii (about Reigate).
*Lychnis diurna.
*Malva moschata.
Mentha Pulegium.
*Menyanthes trifoliata.
Mœnchia erecta.
Myosurus minimus (W. end of the Park).
Myriophyllum spicatum (Earlswood Common).
*Orchis maculata.
*Oxalis Acetosella.
*Potamogeton polygonifolius.
*Radiola Millegrana.
*Rhamnus Frangula (swampy copses).
Saxifraga granulata (lane W. end of Reigate Park, and on the Redhill road).
*Scirpus sylvestris.
Spiranthes autumnalis.
*Stellaria glauca.
Teesdalia nudicaulis.
*Viola palustris.

CRYPTOGAMS.

*Athyrium Filix-fœmina (swampy wood, W. end).
*Dicranella cerviculata.
*Hypnum cordifolium, stramineum; albicans, &c.
Jungermannia platyphylla, &c.
Lomaria Spicant.

Nephrodium dilatatum; *spinulosum.
*Pilularia globulifera (Earlswood Common).
Sphagnum cymbifolium; squarrosum (swampy wood).

Note.—In Cooper's 'Flora Metropolitana,' 1837, mention is made in the Supplement of a bog on the Redhill road, two miles beyond Merstham; there is no reference to it in the Floras of Surrey and Reigate. Comarum, Schœnus, and Blysmus, &c., were said to grow there. The bog and adjoining fields have probably long since been drained, to the extinction of the plants.

ADDENDA.

Lane between Reigate Church and Ffrenches, Mentha rotundifolia; meadows left of the Dorking road, Orchis latiflolia; meadows, Hightrees Farm, &c., Orchis Morio; Colley woods, Carex pendula; Chart Lane, Aspidium aculeatum, angulare; about Reigate, (lanes ?) Asplenium Trichomanes; lane E. of Reigate and between Ffrenches and Luckfield Street Saponaria officinalis.

42. MERSTHAM, AND HILLS EAST OF MERSTHAM, REDSTONE HILL.

The hills east of Merstham are of the same character as those described in preceding localities which treat of the chalk range. White Hill stands out as most prominent in the section. Here also are extensive excavations, and the chalk grit and refuse at the base of the hill is prolific in plants peculiar to the formation. The beechwoods above, and copses on the flanks of these hills, especially in the direction of Caterham, are also good, and apparently unfrequented localities. Between the downs and Redstone Hill, parallel to them, are some marshy meadows and swampy corners; near Merstham, some pools; Redstone Hill rises due east of Redhill; a road to Nutfield and Bletchingley runs along its undulating heights. The slopes are cultivated and copses mostly enclosed.

CHALK HILLS.

*Anthyllis Vulneraria.
*Asperula cynanchica.
Atropa Belladonna (between Merstham and Godstone).
*Avena pubescens (plentiful).
*Brachypodium pinnatum.
*Briza media.
*Bromus erectus.
*Calamintha Clinopodium.
*Campanula Trachelium.
*Carduus acaulis.
*Carlina vulgaris.
*Cephalanthera grandiflora.
*Chlora perfoliata.
*Cistus Helianthemum.
*Clematis Vitalba.
Conium maculatum.
*Cornus sanguinea.
*Cynoglossum montanum.
*Daphne Laureola.
*Echium vulgare.
*Epipactis latifolia (beech groves abundant towards Caterham).
*Erigeron acris.
*Erythræa Centaurium.
Euphorbia amygdaloides; platyphylla.
*Euphrasia officinalis.
*Gentiana Amarella (plentiful on the slope towards Caterham Valley).
Habenaria chlorantha; bifolia.
*Hippocrepis comosa.
Hypericum Androsæmum;* hirsutum montanum.
*Inula Conyza.
*Juniperus communis.
*Kœleria cristata.
*Lathyrus sylvestris (abundant).
*Linum catharticum.

*Lotus corniculatus.
Ophrys aranifera.
*Orchis pyramidalis.
*Origanum vulgare.
Paris quadrifolia (in copses).
*Picris hieracioides.
*Poterium Sanguisorba.
*Primula vulgaris (woods).
*Reseda lutea; Luteola.
*Rosa micrantha; *rubiginosa.
Scabiosa arvensis; Columbaria (abundant about Caterham).
*Senecio erucifolius; *sylvaticus.
*Taxus pinnatus.
*Viola hirta.

Borders of pools and swampy meadows about Merstham and east of it.

Cardamine amara.
Carex pulicaris; paniculata; pendula; pallescens.
Chrysosplenium oppositifolium; alternifolium.
Epipactis palustris.
Eupatorium cannabinum.
*Lathyrus sylvestris (adjoining banks, &c.).
Orchis latifolia.
*Silaus pratensis.
*Viola palustris.

REDSTONE HILL.

Copses: Euphorbia amygdaloides; Hieracium vulgatum; Solidago Virgaurea; Dianthus Armeria. Borders of rill and marshes about Ham Farm:* Stellaria aquatica; Blysmus compressus (?). Lane leading from Redstone Hill towards the downs:* Calamintha menthifolia;* Hieracium vulgatum; Viola odorifera; Bryum hornum, &c.; Scolopendrium vulgare.

43. THE BETCHWORTH HILLS.

The river Mole, which drains the country immediately south of the downs between Dorking and the hills east of Merstham, pursues its course northwards to the Thames, through an opening in the range opposite Dorking. Here are the Betchworth Hills; the most westerly of which, fronting Betchworth Park on one side, and on another side the valley of the Mole, is Box Hill—well known to collectors and excursionists, and so called from the Box plant which grows upon its precipitous flank in the greatest profusion. On the opposite side of the valley is Ranmer Common, which attains an altitude equal to that of Box Hill, and at Mickleham another valley with a road along it to Headley extends in an easterly direction, and separates the hill from the heights known as Mickleham Downs. A wood of considerable extent covers the upper part of Box Hill, but the plateau above the Betchworth Hills proper is more open, and is continuous with Headley and Walton Heaths; while their flanks, chalky slopes, are more or less covered with Juniper, and excavated in places. Subsoil above, gravel. Meadows by the Mole are marshy, at any rate damp.

Aceras anthropophora.
*Ajuga Chamæpitys (cornfield above).
*Anthyllis Vulneraria.
Arabis hirsuta.
*Asperula cynanchica; odorata.
Astragalus Glycyphyllos (Headley Lane).
*Atropa Belladonna.
*Brachypodium pinnatum.
*Briza media.
*Bromus erectus (opposite Betchworth).
Bunium flexuosum.
*Buxus sempervirens (abundant).
*Campanula glomerata; *rotundifolia; Rapunculus; *Trachelium.
*Carduus acaulis; *nutans (fields above; plenty).
*Carlina vulgaris (Betchworth Downs).
Caucalis daucoides.
*Cephalanthera grandiflora (woods); ensifolia (woods lef of Headley Lane).
*Chlora perfoliata.
*Circæa lutetiana (woods).
*Cistus Helianthemum.
*Clematis Vitalba.

*Clinopodium Calamintha; *Nepeta (bank behind Burford Bridge station); *Acinos (fields).
*Conium maculatum (in the hollow right of Headley Lane).
*Cornus sanguinea.
Crepis fœtida (Juniper Hill).
Cuscuta europæa (on nettles by the Mole).
*Cynoglossum officinale (in the woods behind Mickleham Church); montanum ? (bank near Mickleham).
*Daphne Laureola; Mezereum.
Digitalis purpurea (woods).
*Echium vulgare.
*Epilobium angustifolium (Box Hill, above plentiful).
*Epipactis latifolia (abundant, woods); palustris.
Erigeron acris.
*Erythræa Centaurium.
*Euphrasia officinalis.
*Galeopsis Ladanum (fields; also in hollow right of Headley Lane).
Genista tinctoria ?
*Gentiana Amarella.
Geranium columbinum; *Phæum (road-side, Mickleham); pratense.
Gymnadenia conopsea.
Habenaria chlorantha; bifolia.
Helleborus viridis (copse near Mickleham); fœtidus (wooded slope left of Headley Lane).
*Herminium Monorchis (slope of hollow right of Headley Lane).
Hesperus matronalis (copse near Mickleham).
*Hippocrepis comosa.
*Hypericum hirsutum; *calycinum (banks about Mickleham).
*Inula Conyza.
*Juniperus communis.
Lathræa Squamaria (copse, Mickleham).
Lepidium Smithii.
*Linum catharticum; angustifolium.
*Listera ovata.
*Lotus corniculatus.
Lysimachia nemorum.
Medicago maculata; lupulina.
*Monotropa Hypopitys.
Neottia Nidus-avis.
*Ononis arvensis.
*Ophrys apifera (slopes of hollow right of Headley Lane); muscifera.
Orchis mascula; *pyramidalis; ustulata; maculata; Morio.
*Origanum vulgare.
Orobanche elatior.
Papaver Argemone.
Phyteuma orbicularis.
*Polygala vulgaris; calcarea.
Poterium Sanguisorba.
*Pyrus torminalis.
*Reseda lutea; Luteola.
*Rosa micrantha.
*Rubus Idæus.
Ruscus aculeatus.
Sagina nodosa ? (grassy slopes).
Salix nigricans ? (foot of Box Hill).
Scandix Pecten-Veneris (fields).
*Scleranthus annuus (cornfield).
Sedum acre.
*Senecio erucifolius; *sylvaticus.
Specularia hybrida (cornfield).
Spiræa Filipendula.
Spiranthes autumnalis.
Tanacetum vulgare (Burford Bridge).
*Taxus baccata.
*Teucrium Botrys (hollow, right of lane leading to Headley; plenty, 1874).
Thesium linophyllum.
*Torilis infesta.
Valerianella dentata (fields).
*Verbascum nigrum; Thapsus; Thapso-nigrum.
*Verbena officinalis.
*Viburnum Lantana; Opulus.
Vicia Cracca.
*Viola hirta; sylvestris; tricolor; *odorata (below).

Cryptogams.

Ceterach officinarum (Headley Lane (?) olim).
Cryphæa heterophylla (foot of Box Hill).
Cylindrothecium concinnum.
*Homalothecium sericeum (walls near Burford Bridge).
Hypnum crassinervium; pumilum; *splendens; brevirostre; *triquetrum; *loreum; molluscum.
Leptotrichum filiforme (slopes of Box Hill).
*Leucodon sciuroides.
Orthotrichum leiocarpum; Lyellii.
Thamnium alopecurum.
Thuidium tamariscinum; abietinum, &c.

Addenda.

Mickleham and Leatherhead Downs:—

*Brachypodium pinnatum.
*Campanula glomerata.
*Carduus acaulis.
*Carex glauca.
*Cephalanthera grandiflora (copses).
*Erythræa Centaurium.
*Gentiana Amarella.
*Helianthemum vulgare (in large patches).

Melampyrum cristatum (?) (wood near Headley).
Orchis ustulata; *pyramidalis.
Spiræa Filipendula.
Spiranthes autumnalis.
Thesium humifusum.

44. ABOUT BUCKLAND AND BROCKHAM.

The downs between Betchworth and Reigate hereabouts, are apparently of lower elevation than the sections right and left of them, though actually the difference is not great. This arises from the extent of the cultivated slopes at their base and the more gradual rise; otherwise the topographical character of the locality is much the same. Brockham village is on the banks of the river Mole, which approaches the downs at this point.

*Anthyllis Vulneraria.
*Asperula cynanchica.
*Avena pratensis (foot of the downs); *pubescens.
*Bromus erectus.
*Calamintha Clinopodium (banks and fallows).
*Carduus acaulis.
*Carlina vulgaris.
*Chlora perfoliata.
*Crepis taraxacifolia (banks and fallows).
*Erigeron acris.
*Erythræa Centaurium.
*Geranium Columbinum (copses).
*Hypericum montanum.
*Inula Conyza (banks and fallows).
Kœleria cristata.
*Ophrys apifera (hills opposite Brockham).
*Orchis pyramidalis; *ustulata.
*Picris hieracioides.
*Reseda lutea.
*Scabiosa arvensis, &c.

Cornfields and Fallows.

Alchemilla vulgaris (meadows near Dorking).
*Ajuga Chamæpitys.
*Anagallis cærulea.
Barbarea præcox.
Bupleurum rotundifolium (near the Hermitage).
Calamagrostis Epigejos (hedgebanks).
*Calamintha Acinos; *menthifolia.
*Campanula Trachelium.
Caucalis daucoides (fallows above Buckland Hill).
Crepis biennis (fields near).
*Euphorbia platyphylla.
Galanthus nivalis (meadows towards Betchworth).
*Galeopsis Ladanum.
*Galium tricorne.
Hieracium boreale (bank near Reigate road).
Leersia oryzoides (by Brockham Bridge and elsewhere near).
*Linaria spuria; Elatine; *minor.
*Mentha arvensis.
*Nuphar; *Tanacetum; *Potamogeton densus (and other common waterside plants by the river); Fumaria capreolata (agg. between Brockham and the downs).
Ophioglossum vulgatum (meadows).
Potentilla argentea (lane near).
Ranunculus parviflorus &c. (copses above Brockham Hill).
Salix rubra (var.).
*Stachys arvensis (near Brockham).
*Thlaspi arvense.
*Viburnum Lantana &c. (abt. Brockham).
Viscum album (Betchworth Park).

45. RANMER COMMON, AND THE HILLS WEST OF DORKING AND WHITE DOWNS.

Ranmer Common, on the summit of the hills, on the other side of the Mole, can be approached either from the Dorking station on the Redhill and Guildford line, or from the Burford Bridge station, whence Westhumble Lane leads to it directly. It lies high, and is surrounded by woods. The ascent from Westhumble Lane by its northern flank is well covered with scrub and brushwood. Immediately facing Box Hill are the grounds of Mr. Cubitt's mansion. The common extends behind these grounds, and below them, facing Dorking, are some old chalk quarries.

Westwards the range is continued uninterruptedly as far as Guildford, with a general similarity in its features to those of the sections already described—grassy downs sprinkled with Juniper, surmounted by frequently recurring beechwoods, with slopes below fallow, or in a state of cultivation. In the vale below, a rivulet with pasturage; beyond these the wooded rising ground at the base of the red sandstone hills of the Wealden, which in the neighbourhood of Dorking attain an elevation of nearly one thousand feet. East of Wotton, the drainage falls into the Mole; west of that place it is tributary to the Wey. The subsoil of the common is gravel.

COMMON; BORDERING WOODS, ETC.

*Aquilegia vulgaris (Westhumble Lane).
Arabis hirsuta (Ashdown Copse).
Atropa Belladonna (woods about Denbies).
*Campanula Trachelium (Westhumble Lane, plentiful).
*Chelidonium majus (Westhumble Lane).
*Digitalis purpurea.
Galeopsis Ladanum (field near).
Gnaphalium sylvaticum.
Helleborus viridis (copse near).
Hordeum sylvaticum (copse behind the 'Fox').
Hypericum montanum; hirsutum.
*Lamium Galeobdolon.
*Lathræa Squamaria (Westhumble Lane in a hazel copse; also roadside).
Lithospermum officinale (copses).
Neottia Nidus-avis (woods).
Ophrys muscifera (near Denbies).
Primula vulgaris (copses); *veris (fields left of the lane).
*Ranunculus auricomus (lane).
*Rosa rubiginosa; *micrantha.
Ruscus aculeatus (Westhumble Lane and Ashdown Copse).
Thesium linophyllum (roadside half-mile beyond the chalk quarries and Ranmer Hills).
*Viola odorata.

CHALK QUARRIES.

These quarries are partly new, and partly old; they are of considerable extent, and are situated near the railway station.

Aceras anthropophora (?).
*Anthyllis Vulneraria.
Briza media.
*Bromus erectus.
*Calamintha Clinopodium.
*Campanula glomerata.
*Carduus acaulis.
*Centaurea Scabiosa.
*Chlora perfoliata (near).
*Cichorium Intybus.
*Clematis Vitalba.
Crepis fœtida.
*Daucus Carota.
*Echium vulgare.
Galeopsis Ladanum (fields above); ochroleuca (? olim).
*Hieracium murorum.
*Hippocrepis comosa.
*Inula Conyza.
*Ononis arvensis.
Ophrys apifera (?); muscifera (?).
Orchis pyramidalis.
*Origanum vulgare.
Orobanche minor (clover field near).
*Phyteuma orbiculare.
*Picris hieracioides.
*Poterium Sanguisorba.
Primula veris; vulgaris.
*Reseda lutea.
*Scabiosa Columbaria.
Silene inflata.
Smyrnium Olusatrum.
*Tanacetum vulgare.
*Torilis infesta (fields near).

HILLS WEST OF DORKING.

Ajuga Chamæpitys (fields).
*Anthyllis Vulneraria.
*Asperula cynanchica.
*Briza media.
*Bromus erectus.
*Calamintha Clinopodium; *Acinos.
*Campanula glomerata.
*Carduus acaulis; tenuiflorus.
*Carlina vulgaris.
*Centaurea Scabiosa.

*Cichorium Intybus.
*Daucus Carota.
*Echium vulgare.
*Erigeron acris.
*Erythræa Centaurium.
*Euphrasia officinalis.
*Galeopsis Ladanum (fields).
*Gentiana Amarella.
*Helianthemum vulgare.
*Hypericum hirsutum.
*Inula Conyza.
*Juniperus communis.
*Linum catharticum.
*Orchis pyramidalis.
*Origanum vulgare.
*Pastinaca sativa.
*Picris hieracioides.
*Reseda lutea; *Luteola.
*Rosa micrantha.
*Scabiosa arvensis; *Columbaria.
*Taxus communis.
*Thymus Serpyllum.
*Verbascum Thapsus.
*Verbena officinalis.
*Viburnum Lantana.

Note.—A deep gully with steep sides nearly opposite Westgate, should be examined together with bordering woods.

APPENDIX.

In Norbury Park, Neottia Nidus-avis; Herminium Monorchis; Ophrys muscifera; Epipactis grandiflora; Dipsacus pilosus; in Beeching Wood, Cynoglossum montanum.

Guildford road, near Effingham chalk-pit, on a bank: Thesium linophyllum; Phyteuma orbiculare.

Fetcham Downs, Bagdon Hill near Bookham: Teucrium Botrys?

46. GUILDFORD AND THE HILLS EAST OF GUILDFORD.

Here another break occurs in the chalk downs, through which the river Wey joined by the Sittingbourne stream, which drains the country east of Guildford and Godalming, pursues its course to the Thames at Weybridge. A peculiar feature of the range at this point, is the narrowness of the ridge to which it has gradually been contracted. Fronting them to the southward are the redsand hills of the Wealden, of which an isolated parallel ridge runs for a short distance in the same direction, and for an interval separates the chalk downs from the rivulet above mentioned; but further eastwards towards Albury this stream with its bordering meadows lies at the foot of the downs; nearest to Guildford are the Merrow Downs; further on and above Albury are the Clandon Downs; beyond these is Netley Heath, with Shiere and Gomshall in the vale below. The topographical features of this are similar to those of other sections of the chalk range already described, and a general similarity will be found to prevail with respect to its vegetation. Beyond Guildford are extensive chalk-pits, both old and recent; and near these is the village of Shalford, below the downs and on the Godalming road. An open grassy common here, with two ponds upon it.

Aceras anthropophora (downs).
Actinocarpus Damasonium (Shalford Common)? olim.
*Aira caryophyllacea.
*Alchemilla arvensis (downs).
Allium vineale (hill above Vale's cottage, Shiere).
Anemone apennina (woods about Guildford).
*Anthyllis Vulneraria (chalk-pits).
Aquilegia vulgaris (Netley Wood); Shiere.
*Arabis hirsuta (Guildford chalk-pits,

scarce); perfoliata (lanes, Shiere and Albury).
*Avena flavescens (downs); *pubescens.
Barbarea præcox (banks of the Wey).
*Brachypodium pinnatum (downs).
Bupleurum rotundifolium (cornfields, brow of hill, over against St. Martha's Chapel, towards the Merrow Downs).
Calamagrostis Epigejos (Weston Wood, Albury).
*Calamintha Clinopodium (chalk-pits); *Acinos (fields); menthifolia (St. Catherine's Hill).
*Carduus nutans (also on Netley Heath); *acaulis.
*Centaurea Scabiosa.
Chrysosplenium oppositifolium (Alder Copse, Wood Farm and elsewhere; about Guildford; between Shalford and St. Martha's Chapel, in moist copses).
*Cistus Helianthemum.
*Clematis Vitalba.
Conium maculatum (swampy places, about Albury and Shiere).
Corydalis claviculata.
*Crepis taraxacifolia (chalky fields).
Cuscuta Trifolii (about Guildford and St. Martha's Chapel); europæa (osier holt).
Cynoglossum officinale (Guildford).
*Cyperus fuscus (pond, Shalford Common).
Dianthus Armeria (footpath between Guildford and Albury).
Digitaria humifusa (? about St. Martha's Chapel; olim).
*Diplotaxis tenuifolia (chalk-pits).
Dipsacus pilosus (near Chilworth, &c.).
Echinochloa Crus-galli (about St. Martha's Chapel, in arable land).
*Echium vulgare (pits).
Epilobium roseum (near Albury Church, by the rivulet; about Shiere).
Epipactis latifolia (woods about Shiere).
Eranthis hyemalis (Albury Park).
*Filago germanica (downs); gallica (cornfield between Chilworth road and St. Martha's Hill).
*Fumaria parviflora.
Galanthus nivalis (Stoke Park).
Galeopsis Ladanum.
Geranium columbinum (Losely).
Helminthia echioides.
Hieracium murorum and boreale (about Albury and Shiere).
Hippuris vulgaris (pond, Clandon Park).
Hyoscyamus niger (Merrow Downs, Gomshall; not on Shalford Common).
Hypericum Androsæmum (Weston Wood, Albury and Stoke Wood); montanum (Shiere).
Impatiens fulva (bank of the Wey and of the Sittingbourne).
Inula Conyza.
*Isatis tinctoria (chalk-pits, and fields about them).
*Juniperus communis (downs).
Kœleria cristata (downs; Shiere).
Lathræa Squamaria (field between Chantry Downs and Shalford turnpike).
*Lepidium campestre.
Limosella aquatica (Shalford Common).
*Linum catharticum (pits, &c.).
Myosotis cæspitosa (Shalford Common).
Neottia Nidus-avis (Netley Wood, Shiere, and between Guildford and St. Martha's Chapel).
Ophrys apifera (about Shiere and Albury).
*Orchis pyramidalis (downs).
Origanum vulgare.
*Papaver hybridum (cornfields).
*Parietaria diffusa (Shalford).
Phyteuma orbiculare (downs about Albury, Shiere, and Gomshall).
*Picris hieracioides (chalk-pits).
Pimpinella magna (about Guildford).
Polygonum Bistorta (Wotton meadows).
Potentilla argentea (St. Martha's Hill).
*Poterium Sanguisorba.
Prunus Padus (about Shiere and Gomshall).
*Reseda lutea.
Sagina nodosa (Shiere and Albury).
Salvia verbenaca.
Saponaria officinalis (St. Martha's Hill, also near Shiere).
Saxifraga granulata (St. Catherine's Hill and about Shiere.)
*Scabiosa arvensis: *Columbaria.
*Scandix Pecten-Veneris.
*Scirpus acicularis (Shalford Common).
*Scleranthus annuus; perennis.
Sedum Telephium (Albury).
Setaria viridis (between Guildford and Albury).
*Specularia hybrida (cornfields).
*Spiræa Filipendula (downs).
Stellaria glauca (meadows by the Wey and below St. Martha's Hill).
Teesdalia nudicaulis (Albury and Shiere heaths).
Thesium humifusum (downs near).
Trifolium subterraneum (St. Martha's Hill); scabrum (Netley Heath).
Valerianella Auricula (between Guildford and St. Martha's Hill).
*Verbascum nigrum (between Guildford and Shalford; no other species there now).
Veronica montana (about Shiere and Albury).
Vinca minor (Albury).

Cryptogams.

Asplenium Adiantum-nigrum (woods and lanes about Albury and Shiere and

Gomshall); Trichomanes (near Shiere); Ruta-muraria (Merrow Downs and about Shiere).
Cystopteris fragilis (low wall west of the road from Weston Street to Albury Park ? olim).
Nephrodium spinulosum (bogs near Shiere); dilatatum (Albury).
Ophioglossum vulgatum (meadows in Albury Park; Losely Park near Guildford).

MOSSES ABOUT SHIERE, GOMSHALL, ETC., FROM 'SCIENCE GOSSIP.'

Hypnum illecebrum; irriguum; loreum; polymorphum; stramineum.
Leptodon Smithii.
Mnium rostratum.
Neckera crispa; pumila.
Orthotrichum anomalum; Sprucei; pulchellum; stramineum; tenellum.
Physcomitrion pyriforme.
Pottia truncata.
Pterygonium gracile.
Schistostega pinnata.
Tetraphis pellucida.
Thuidium tamariscinum.
Tortula vinealis; Hornschuchiana; cuneifolia; marginata; latifolia.
Trichostomum crispulum.
Webera albicans.
Weissia calcarea.

47. WHITEMOOR COMMON.

This common, three miles north of Guildford, is an extensive heathy tract, low lying in the centre and elevated westwards and eastwards. A road from Guildford to Chobham passes by the village of Worplesdon at its south-western extremity, and a branch of the South-Western Railway intersects it from north to south. The large pond which stood formerly upon the heath has been drained, and converted into pasturage; subsoil sand, more or less peaty; central parts marshy, slopes in several places, boggy. A few Fir-trees have been planted at one end of the common, otherwise the vegetation consists of Furze, Ling and Bracken, with the following plants.

*Achillea Ptarmica.
*Anagallis tenella.
*Anthemis nobilis.
*Bidens tripartita; *cernua (ditches near).
*Carduus pratensis.
*Cuscuta Epithymum.
*Drosera rotundifolia; *intermedia.
*Erica cinerea; Tetralix.
*Eriophoron angustifolium.
Gentiana Pneumonanthe.
*Helosciadium inundatum.
*Hydrocotyle vulgaris.
*Hypericum Elodes (bogs).
*Inula Pulicaria.
*Juncus squarrosus, &c.
*Littorella lacustris.
*Lychnis Flos-cuculi.
*Mentha Pulegium (near Worplesdon); *sativa; *pratensis (between Worplesdon and Pirbright, by a rill).
*Molinia cærulea.
*Narthecium ossifragum (abundant).
*Œnanthe fistulosa.
*Potamogeton polygonifolius.
*Radiola Millegrana.
*Rhynchospora alba (abundant).
*Salix repens.
*Scirpus cæspitosus.
*Scutellaria galericulata.
*Solidago Virgaurea (dry parts, eastwards).
*Triodia decumbens.
*Ulex Gallii.

CRYPTOGAMS.

*Aulocomnion palustre.
*Lycopodium inundatum (abundant).
*Sphagnum cymbifolium, &c.

48. BISLEY COMMON, PIRBRIGHT HEATH, AND COW MOOR.

Westwards, and distant two miles from the Worplesdon referred to in the preceding locality, is the village of Pirbright; surrounded, with the exception of a grassy flat and a few patches of cultivated land, by vast tracts of heath and moor land; hilly in parts, and rising westwards to the elevated steeps known as the Fox Hills and Chobham Ridges. The Farnham branch of the South-Western Railway intersects it, and separates Pirbright Heath from the moors and common. In many places Fir-trees have been planted, otherwise the vegetation consists of Heath, Furze and Bracken; the Furze frequently disappearing for a space; bogs many; subsoil sand, often peaty; of the hills, sand and gravel.

*Aira flexuosa; *caryophyllacea; cæspitosa (moors).
*Alisma ranunculoides (by the canal).
*Anagallis tenella (bogs).
*Carduus pratensis.
*Carex binervis (heaths); vulgaris; *ampullacea (bogs).
*Ceratophyllum aquaticum.
*Cuscuta Epithymum.
*Digitalis purpurea.
*Drosera rotundifolia; *intermedia.
Elatine Hydropiper (small pond on the heath near Pirbright).
*Epilobium palustre; *parviflorum.
*Erica cinerea; *Tetralix.
*Erigeron acris (banks of canal).
*Eriophorum angustifolium; *vaginatum.
*Euphrasia officinalis.
Gentiana Pneumonanthe (east of the cemetery towards Woking).
*Hieracium vulgatum.
*Hydrocotyle vulgaris.
*Hypericum Elodes.
*Inula Pulicaria (about Pirbright).
*Juncus squarrosus; *acutiflorus, *lamprocarpus, &c.
*Littorella lacustris (pond near the railway and by the canal).
*Lycopus europæus (canal).
*Lysimachia vulgaris (ditches about Pirbright village).
*Molinia cærulea.
*Myrica Gale (moory parts, abundant).
*Narthecium ossifragum.
*Pedicularis palustris (by the canal).
*Polygonum minus (Pirbright Green).
*Potamogeton heterophyllus (canal); *polygonifolius (bogs); lucens (canal).
*Rhynchospora alba (Cow Moor, plenty).
*Radiola Millegrana (wet sandy places).
*Rubus carpinifolius (by the canal).
Sagina nodosa (banks of the canal, near Pirbright).
*Salix repens; *cinerea.
*Scirpus multicaulis; *cæspitosus; acicularis.
*Scutellaria galericulata; *minor.
*Solidago Virgaurea.
*Teucrium Scorodonia.
*Triodia decumbens.
*Ulex Gallii.
*Utricularia vulgaris (ditches and holes, by a pond near the railway).
*Vaccinium Myrtillus.
*Verbascum Thapsus.

CRYPTOGAMS.

*Aulocomnion palustre.
*Lomaria Spicant.
*Lycopodium inundatum.
*Nephrodium Filix-mas.
*Sphagnum cymbifolium; *acutifolium.
Etc.

49. WOKING HEATH, HORSELL COMMON, AND BANKS OF THE BASINGSTOKE CANAL.

Enclosures and buildings, more or less continuous, have here diminished the area of what was once a very considerable tract of heath, in the neighbourhood of Woking Common station. It was continuous with that portion of Pirbright Common now included in the precincts of the Woking necropolis. The Basingstoke Canal and the South-Western Railway intersect it, from east to west, with plantations of Fir-trees on either side of them, more or less continuous as far as Weybridge. Horsell Common and Woodham Heath (one expanse) adjoin it; the former crossed by a road to Chobham, and the latter bordered by a road to Chertsey. Subsoil sand, occasionally peaty.

BY THE CANAL.

*Achillea Ptarmica.
*Alisma ranunculoides.
*Eriophorum angustifolium.
*Hypericum Elodes; *pulchrum.
Leersia oryzoides (by the bridge over the canal).
*Littorella lacustris (abundant).
*Lotus major.

*Lycopus europæus.
*Lysimachia vulgaris; *Nummularia.
Sagina nodosa.
*Sagittaria sagittifolia, &c.
*Scirpus palustris; *multicaulis; *acicularis (Milford Green, plentiful by the pond).
*Scutellaria galericulata.
*Stachys palustris.
*Stellaria glauca.

ON THE COMMONS.

*Anagallis tenella.
*Calluna and *Erica.
*Euphrasia officinalis.
Gentiana Pneumonanthe (? olim near the station).
*Hieracium vulgatum; *umbellatum.
*Molinia cærulea.
*Nardus stricta.
*Potamogeton polygonifolius.
Sagina ciliata (west of the station).
*Salix repens.
*Scirpus cæspitosa.
*Solidago Virgaurea (Horsell Common, plenty).
*Ulex Gallii.

CRYPTOGAMS.

*Lomaria Spicant.
Lycopodium inundatum (other species formerly near the station).
*Pteris aquilina.
*Scolopendrium vulgare.
*Sphagnum cymbifolium, &c.

50. WEYBRIDGE, ST. GEORGE'S HILL, AND BANKS OF THE THAMES ABOUT WALTON.

The sandy fields about Weybridge were formerly the best localities near the metropolis for some uncommon plants, affecting light sandy soil; but they are now rare, if not extinct. The sand-hills south of the railway line from Esher, thus far, attain their greatest elevation at Weybridge; they are known as St. George's Hills, are planted with Fir and other trees, and enclosed, but open to the public as a promenade. Weybridge itself, together with some patches of heath still unenclosed, is also surrounded on all sides, but that of the Thames, by plantations of Fir. South of St. George's Hills in the direction of Cobham are tracts of furze-grown commons, which are crossed by the Guildford road.

*Aira flexuosa (Weybridge).
*Antirrhinum Orontium (cornfields, Cobham, &c.).
Arnoseris pusilla (?) (Weybridge, olim; also between Hersham and St. George's Hill, in sandy fields).
*Butomus umbellatus (banks of the Wey).
Campanula Rapunculus (between Cobham and Stoke).
Carex elongata (meadows between canal and river).
*Carex præcox (roadside bank).
Claytonia perfoliata (hedgebanks on the common).
*Digitalis purpurea.
Digitaria humifusa (fields, olim).
*Diplotaxis tenuifolia (about the church walls).
*Epilobium angustifolium (railway banks).
*Erica cinerea.
*Erysimum cheiranthoides (fields).
Filago apiculata (several places).
*Fumaria officinalis.
Genista anglica.
Gnaphalium sylvaticum (St. George's Hill, and on Fairmile Common).
*Hydrocotyle vulgaris (pond on St. George's Hill).
*Hypericum Elodes (pond on St. George's Hill).
Hypochœris glabra (fields, olim).
*Hieracium umbellatum; vulgatum; tridentatum (Fairmile Common).
*Leontodon hirtus; *hispida.
*Lysimachia vulgaris (by the Wey).
*Lythrum Salicaria (by the Wey).
*Œnothera biennis (railway banks).
Rhynchospora alba (swamp near Cobham).
Sagina ciliata.
*Sarothamnus scoparius.
*Senecio sylvatica.
Setaria viridis (fields, olim).
*Tanacetum vulgare (banks of the Wey).
Teesdalia nudicaulis (between Hersham and Weybridge).
*Teucrium Scorodonia.
*Ulex Gallii (furzy commons).

BANKS OF THE THAMES.

Barbarea vulgaris.
*Butomus umbellatus.
*Diplotaxis tenuifolia (frequent).

*Erysimum cheiranthoides.
*Humulus Lupulus (hedges).
*Hydrocharis Morsus-ranæ (at Walton).
*Limnanthemum nymphæoides (at Walton in a backwater, abundant).
*Lysimachia vulgaris.
*Myriophyllum verticillatum (at Walton).
*Nuphar lutea.
*Potamogeton perfoliatus.
*Ranunculus fluitans.
*Rhamnus catharticus (hedges).
*Rumex Hydrolapathum.
*Sagittaria sagittifolia.
*Salix vitellina.
*Sium latifolium (two or three plants).
*Thalictrum flavum.
*Trifolium fragiferum.
*Viburnum Opulus (a shrub or two).

CRYPTOGAMS.

*Chara vulgaris (ditch opposite Sunbury).
*Fontinalis antipyretica (on stones).

51. CHOBHAM COMMON.

Westwards of Weybridge and Chertsey, and right of the Guildford Railway, are the wide tracts of Chobham and Bagshot, continuous with each other, and broken only in their continuity by some reclaimed and enclosed land about Chobham and Wendlesham, watered by the Bourne rivulet which drains the slopes of the sand-hills, and of the uplands beyond Bagshot. On these slopes in many places there are patches of bog and marsh, which drain, generally speaking, into ponds in the hollows below. Glover's Ponds have been wholly, and Gracious Pond partly drained; otherwise the whole district may be considered an undulating waste of sand and heath, extending over many thousand square acres of country, and imparting a peculiar character to this corner of Surrey. A road leads direct from Chertsey to the common, and across it to Chobham and Bagshot; distance, three miles; the nearest point to the west common is Woking station.

*Alnus glutinosa (Gracious Pond.)
*Anagallis tenella (bogs).
*Antirrhinum Orontium (cornfield, Chobham).
*Aquilegia vulgaris (roadside bank, near Long Cross).
*Bidens tripartita (ditches, Chertsey).
*Carduus pratensis.
*Carex flava (marshy places).
*Circæa lutetiana (ditches, Chertsey).
*Chrysanthemum segetum (cornfield, Chobham).
*Cuscuta Epithymum.
*Digitalis purpurea (bordering hedges and copses).
*Drosera rotundifolia; *intermedia.
*Erica cinerea; Tetralix.
*Eriophorum angustifolium; *vaginatum.
*Erysimum cheiranthoides (fields, Long Cross).
*Euphrasia officinalis.
*Genista anglica.
*Gentiana Pneumonanthe (near an isolated clump of Fir-trees towards Long Cross, plentiful).
*Hieracium umbellatum; *vulgatum.
*Hydrocotyle vulgaris.
*Hypericum Elodes.
Hypochœris glabra (field near Chobham).
*Juncus squarrosus; *acutiflorus; *glomeratus, &c. (drained ponds).
*Melampyrum pratense (bordering copses).
*Molinia cærulea.
*Nardus stricta.
*Narthecium ossifragum.
*Potamogeton polygonifolius.
*Potentilla Tormentilla.
*Rhynchospora alba.
*Salix repens; cinerea (ponds, &c.).
*Scirpus multicaulis (pond near Long Cross); *palustris; *cæspitosa.
*Senecio sylvatica (heathy borders of the Chertsey road, plentiful).
*Serratula tinctoria (heathy borders of the Chertsey road, plentiful).
*Solidago Virgaurea (heathy borders of the Chertsey road, plentiful).
*Teucrium Scorodonia.
*Triodia decumbens.
*Ulex Gallii.
*Viola odorata (Chertsey road).

CRYPTOGAMS.

*Aulocomnion palustre.
*Dicranum scoparium; *Dicranella.
*Lomaria Spicant.
*Nephrodium dilatatum; spinulosum; *Filix-mas (borders and lanes).
*Polytricha, &c. (hedgebanks).
*Sphagnum cymbifolium.

52. BAGSHOT HEATH.

Bagshot Heath, as stated in the preceding locality, adjoins Chobham Common, and as this consists of two portions separated the one from the other by the village of Chobham and surrounding enclosures, so is Bagshot Heath in a similar manner divided into two parts by the village of Wendlesham with the enclosures and cultivated land thereunto appertaining. Plantations of Fir-trees are more general on Bagshot Heath than on Chobham Common, and on the whole it is the higher lying of the two; the country indicating a sensible rise towards Ascot, as well as towards the ridges behind Bagshot: otherwise they resemble each other in soil and in the general character of their vegetation; each, however, possesses its speciality. The lower portion, west of Wendlesham, is wet and moory. A road crosses obliquely the upper portion of the heath to Sunningdale, where there is a station, and the high road from Staines to Farnham, &c., runs along its upper border; another road runs north and south past Tower Hill, and a third in another direction to Bracknel where there is also a station; Tower Hill is the highest point of a central ridge which crosses this upper portion from east to west. There is not much Furze in the open parts; the vegetation is principally Heath. About Sunningdale is a considerable bog.

Agrostis setacea (ridges westwards of Wendlesham).
*Aira flexuosa; *præcox; *cæspitosa.
Blysmus compressus (Winch. MSS. notes).
Campanula patula (Chobham Lane, near Wendlesham Church).
*Carduus pratensis (bogs).
*Carex binervis; *cæspitosa.
*Centunculus minimus (waste land near Wendlesham, right of the road).
Ceratophyllum aquaticum (Englemoor Pond).
*Digitalis purpurea (bordering copses).
*Drosera rotundifolia; *intermedia.
*Erica Tetralix; *cinerea.
*Eriophorum angustifolium; *vaginatum (Sunningdale bog; both in profusion).
*Erythræa Centaurium.
*Gnaphalium sylvaticum.
*Hieracium vulgatum; *umbellatum.
*Hydrocotyle vulgaris.
*Hypericum Elodes.
*Juncus obtusiflorus; *squarrosus.
*Littorella lacustris (Englemoor pond).
*Lotus corniculatus.
*Molinia cærulea.
*Myrica Gale; (bog, W. of Wendlesham).
*Nardus stricta (moory parts).
*Narthecium ossifragum (bogs).
*Phragmites communis (Englemoor Pond).
*Polygonum minus (about Wendlesham); *Fagopyrum (fields, near Ascot).
*Potamogeton polygonifolius (bogs).
*Radiola Millegrana.
*Rhynchospora alba (bogs).
*Salix repens.
*Sarothamnus scoparius (borders).
*Schœnus nigricans (bog W. of Wendlesham, plentiful).
*Scirpus setaceus; *palustris.
*Scutellaria galericulata; *minor.
*Senecio sylvatica.
*Serratula tinctoria (bordering copses, &c.)
*Solidago Virgaurea, &c.
*Spergula arvensis (cornfields, ridges).
*Teucrium Scorodonia (borders).
*Triodia decumbens (moory parts).
*Vaccinium Myrtillus (ridges towards Frimley).
*Viola palustris (Sunningdale).

CRYPTOGAMS.

*Hair-moss (heaths, &c.); bog mosses.
*Lomaria Spicant (abundant).
*Nephrodia (lanes, Wendlesham).
*Polypodium vulgare (lanes, Wendlesham).
*Pteris aquilina (not abundant).

53. THE HOG'S BACK.

This remarkable formation consists of a narrow chalk ridge, seven or eight miles in length, running from the break in the chain at Guildford due west towards Farnham. The plateau above does not exceed a few hundred feet in width; the slopes on either side are cultivated, with occasionally intervening patches of down on the left flank. The Farnham road traversed the ridge from end to end, but the ascent from Guildford being extremely steep, it has been disused, and a new one constructed which ascends the northern flank by an easy gradient, and merges into the old route a mile or two further on. North of the Hog's Back are the sandy heaths and ridges about Pirbright and Bagshot; south of it are those of the Weald, similar in character, however differing from a geological point of view. The main ridge on this side, at an average distance from the downs of six or seven miles, attains at Hind Head Common an altitude of one thousand feet. Near Puttenham, on one side of the Hog's Back, is an extensive chalk-pit; on the other, Wanborough Wood. An isolated conical hill, called Crookesbury Hill, distant one mile from the western extremity, and to the south of it, is a prominent object in the landscape. The flora is that of the chalk district, generally; the drainage, to the Wey.

*Aceras anthropophora (chalk-pits).
*Anthyllis Vulneraria.
*Asperula cynanchica.
*Briza media.
*Campanula Trachelium (near Guildford)
*Carduus acaulis; *nutans.
*Centaurea Scabiosa (cornfields).
*Chlora perfoliata.
*Clematis Vitalba.
*Cornus sanguinea.
*Crepis taraxacifolia (chalk-pit).
*Daphne Laureola; Mezereum.
*Daucus Carota.
*Echium vulgare (pit).
*Erythræa Centaurium.
*Fumaria parviflora; densiflora.
*Gentiana Amarella.
*Helianthemum vulgare.
Herminium Monorchis.
*Hippocrepis comosa (pit).
*Hypericum hirsutum (pit).
*Ilex aquifolia.
*Inula Conyza.
Lathræa Squamaria (copse).
*Ligustrum vulgare.
*Linaria minor.
Linum catharticum.
Marrubium vulgare.
*Nepeta Cataria.
Ophrys apifera; muscifera.
*Orchis pyramidalis; *maculata; ustulata.
*Origanum vulgare.
*Papaver hybridum (cornfields, plenty).
Phyteuma orbiculare.
*Picris hieracioides.
*Plantago media.
*Reseda lutea; *Luteola.
*Rhamnus catharticus.
*Rosa micrantha.
*Scabiosa arvensis; *Columbaria.
Senecio campestris (near New Inn).
*Taxus communis.
Thesium humifusum.
*Thymus Serpyllum.
*Verbascum Thapsus (near Guildford).
*Verbena officinalis (near Guildford).
*Viburnum Lantana, &c.

CRYPTOGAMS.

Equisetum hyemale (Wanborough Wood).

54. PUTTENHAM, ELSTEAD, AND CROOKESBURY COMMONS.

On Crookesbury Common, the hill included, plantations of Fir-trees are general, and towards Godalming there are some wooded heights; otherwise, the fields excepted which border the river Wey, and a few patches of cultivation round about the various villages, the country immediately south of the Hog's Back is an undulating waste of sandy heaths, distinguished the one portion from the other with difficulty. Those lying nearest to the Hog's Back are named as above.

Puttenham Common.

Hilly and of considerable extent, sloping from north and north-west to south-west; furze- and heath-grown; here and there, a Holly or a Whitethorn bush. A rill from a spring at the foot of the downs feeds a succession of ponds, first in the grounds of Hampton Lodge, west of the heath, and then on the common, and these latter receive the drainage from the upper slopes. Before entering the Lodge precincts the rill forms an alder swamp. The ponds on the common, called Cutmill Ponds, are three in succession, divided from each other by dams; the borders of all are more or less boggy in places; continuous westwards with Crookesbury Common.

*Anagallis tenella (bogs).
*Carex ampullacea; *paniculata (alder swamp, above the Lodge); *paludosa (above the Lodge).
*Comarum palustre (upper pond, near the Alders).
*Drosera rotundifolia; *intermedia; Elatine hexandra; Hydropiper (Cutmill ponds, lowermost).
*Epilobium palustre.
*Erica cinerea; *Tetralix.
*Eriophoron vaginatum.
*Hieracium vulgatum (borders).
Hyoscyamus niger.
*Hypericum Elodes.
Limosella aquatica (ponds).
*Lysimachia vulgaris (alder swamp).
Malaxis paludosa (bog, end of pond next to Hampton Lodge).
*Menyanthes trifoliata.
*Nymphæa alba (pond near the Lodge).
*Papaver Argemone (cornfield at Puttenham, foot of the slopes).
*Potamogeton polygonifolius; *natans.
Pyrola minor (wood near Crookesbury Hill).
*Rhynchospora alba.
Sagina nodosa (Cutmill Ponds).
*Scirpus sylvestris (alder swamp); *setaceus; *acicularis.

Cryptogams.

*Aulocomnium palustre; Sphagnum, &c.
*Bartramia fontana.
*Lomaria Spicant (abundant).
*Lycopodium inundatum.

Hankley or Elstead Common, and Frensham Common.

These are separated from Puttenham Common by the cultivated lands about the river Wey and the village of Elstead; but are continuous in every other direction with a wide undulating waste of heath. In the centre is a ridge running north and south, divided by a depression midway into two portions; heath- and furze-clad. It drains into a large pond, called Stotbridge Pond, connected with which is a bog on one side.

*Aira flexuosa.
Camelina sativa (near Tilford).
*Carex ampullacea; *paniculata.
*Comarum palustre (bog, and by the pond, plentiful).
*Digitalis purpurea (borders).
*Erica cinerea; *Tetralix.
*Eriophorum vaginatum.
*Filago minima.
*Genista anglica.
*Hypericum Elodes (bogs).
*Juncus squarrosus.
*Menyanthes trifoliata.
*Narthecium ossifragum.
*Orchis maculata (swampy places).
*Rhynchospora alba.
*Vaccinium Myrtillus.

Cryptogams.

*Lomaria Spicant.
*Lycopodium inundatum.

55. FRENSHAM COMMON.

A sandy, heathy plain; the two large ponds upon it have gravelly bottoms. Westwards, it is continuous with Farnham Common, of which the greater portion is hilly and planted with Fir-trees. A chalky subsoil reappears in this direction. Abbot's Pond, between this common and the preceding one, has been drained and converted into pasturage.

*Aira flexuosa; *præcox.
Arabis perfoliata (sandy lanes, about).
Arnoseris pusilla.
Campanula patula (sandy lanes).
*Carex arenaria.
*Cuscuta Epithymum.
Elatine hexandra; Hydropiper (ponds).
*Erica cinerea; *Tetralix.
*Littorella lacustris (ponds).
Luzula Forsteri (dry banks, near).
Myosurus minimus (cornfields near Frensham Church).
*Myriophyllum alternifolium (pond).
*Potamogeton polygonifolius, &c.
Potentilla argentea.
Radiola Millegrana.
Ranunculus parviflorus (dry banks).
*Sedum Telephium.
Silene anglica (sandy fields).

Appendix.

Chalky banks, &c., about Farnham: Avena strigosa; Campanula glomerata and Rapunculus; Carex strigosa and stricta; Centunculus minimus; Dipsacus pilosus; Euphorbia platyphylla; Habenaria chlorantha; Hippocrepis comosa; Hyoscyamus niger; Neottia Nidus-avis; Petroselinum segetum; Rhynchospora, and Utricularia vulgaris. Peat-bogs, near, Vaccinium Oxycoccos; spongy bogs, Black Lake, near Cæsar's Camp, Osmunda regalis.

56. WITLEY AND THURSLEY COMMONS.

To the south of Puttenham Common, and a mile or so west of Milford, the first station beyond Godalming, lie the tracts of sandy heath, called Witley and Thursley commons, continuous with each other except in so far as they may be deemed separated by a rivulet which has its origin in some ponds on Witley Common; of these Witley Lagg, or the Forked Pond, so called from its peculiar shape, is the largest; it covers an area of about eleven acres in extent, and receives the drainage of the slopes to the southward.

Witley Common slopes on the north side towards the river Wey, between which and its base lie some wet meadows and moory ground. Thursley Common is less elevated. There are no pools upon it except a millpond at Thursley, but in a depression near the centre, a marshy bottom of some extent. Subsoil, redsand; peaty in places.

*Actinocarpus Damasonium (ditches).
*Aira flexuosa.
Arnoseris pusilla (gravelly field near the Hammer ponds; and between Milford and Witley station).
*Carex paniculata; *ampullacea.
*Chrysosplenium oppositifolium (copse, upper end of the Forked Pond).
Circæa lutetiana (copse, upper end of the Forked Pond).
*Cuscuta Epithymum (abundant).
*Digitalis purpurea (bordering banks).
*Drosera rotundifolia; *intermedia.
Elatine hexandra (millpond, Thursley).
*Erica cinerea; *Tetralix.
*Eriophoron angustifolium.

Filago minima.
*Hydrocotyle vulgaris.
*Hypericum Elodes; *humifusum.
*Juncus obtusiflorus; *lamprocarpus.
*Lysimachia vulgaris.
*Menyanthes trifoliata.
*Molinia cærulea.
Polygonum dumetorum (Witley Lagg).
*Rhamnus Frangula.
Sagina ciliata; nodosa (Hammer ponds).
*Stellaria glauca.
*Teesdalia nudicaulis (on the common).
*Vaccinium Oxycoccos (peat-bog near Borough Farm).
*Wahlenbergia hederacea (Forked Pond, borders).

CRYPTOGAMS.

*Dicranella cerviculata.
*Dicranum scoparium.
*Lomaria Spicant.
*Mnium undulatum (copse, upper end of the Forked Ponds).
*Nephrodium spinulosum.
*Polytricha (three species).
*Sphagna, &c.

57. GODALMING.

This place lies amidst the red-sandstone hills of the Wealden; the greater part of the town in a narrow valley, which, in the more open part towards Guildford, is characterised by marshy flats, through which flows the river Wey. The hill west of this flat is called Hurtmoor.

Adoxa moschatellina.
Aquilegia vulgaris.
Arabis perfoliata (copses).
Astragalus Glycyphyllos (Frith Wood).
Campanula Trachelium.
Cardamine amara.
Carex depauperata (woods, near).
Daphne Mezereum (near Stroud House).
Dianthus Armeria (Frith Hill).
Euonymus europæus.
Fritillaria Meleagris (about Stroud).
Gagea lutea? (meadows near).
Geranium lucidum.
Gnaphalium sylvaticum (slopes towards Hurtmoor).
Hypericum montanum (Frith Hill).
Iris fœtidissima (slopes towards Hurtmoor).
Leonurus Cardiaca (four miles on the road to Haselmere).
Lonicera Xylosteum (hedge near Brook).
Myrica Gale (near Stroud House).
Myriophyllum spicatum (ponds on High Down Heath).
Œnanthe Lachenalii (common meadows).
Ornithogalum pyrenaicum.
Pimpinella magna.
Potentilla argentea (Barnacle Hill, &c.).
Ruscus aculeatus.
Samolus Valerandi (water courses, Hurtmoor).
Stellaria aquatica.
Teesdalia nudicaulis (Milford Heath).
Vaccinium Myrtillus (hills, near).
Veronica montana.

CRYPTOGAMS.

Bartramia pomiformis (hedgebank, road to Haselmere).
Nephrodium Thelypteris (Devil's Punchbowl, Hindhead Common).
Osmunda regalis (Devil's Punchbowl, Hindhead Common).

58. LEITH HILL AND THE HOLMWOOD.

Leith Hill, nearly a thousand feet above the sea-level, lies about four miles to the south of Dorking; but the rise in this direction is gradual, and the slope wooded. The upper part is a sandy, gravelly heath, called Leith Hill Common, and is separated from the lower-lying Holmwood Common north-east of it by Ridland Hill, which is covered with plantations, mostly Fir. The summit can be approached by this route, or direct from Dorking by a shady lane, which leads up the gorge west of Ridland Hill, to the hamlet of Cold Harbour at the foot of the last ascent. There is also a path up the hollow behind Wotton; or if the former route be preferred, the descent can be made either this way, or *via* Abinger to Gomshall, or farther eastwards over the Hurtwood Common, and by Peslik Bottom to Albury. In these hollows are lines of drainage with frequently occurring patches of bog. The Holmwood is a turfy common, much overgrown with a scrub of Briars, Brambles, Furze, Ling, Holly, Sallow, Bracken, White-thorn, and Black-thorn. The frequency of the Holly is a special feature in its aspect. The timber of the wooded flanks other than the plantations of Fir, consists of Oak, Birch, Ash, Holly, Sallows, and an undergrowth of Hazel.

HOLMWOOD COMMON AND WOODS ABOUT.

*Achillea Ptarmica.
Actinocarpus Damasonium.
*Aira cæspitosa (woods).
*Anemone nemorosa.
*Angelica sylvestris.
Barbarea præcox (road from Dorking).
*Betonica officinalis.
*Campanula rotundifolia; *Trachelium.
Carex Œderi (on the common).
*Castanea vulgaris (woods).
*Circæa lutetiana (lanes).
Convallaria majalis (Hurtwood common).
*Digitalis purpurea.
*Epilobium angustifolium (Ridland wood).
*Erythræa Centaurium (Ridland wood).
*Euphrasia officinalis.
*Fragaria vesca.
*Galeopsis Tetrahit (lanes).
Genista tinctoria.
*Hieracium sylvaticum; umbellatum; murorum; boreale.
*Hippuris vulgaris (pond on Bury Hill).
*Hypericum Androsæmum (lanes between Dorking; and Wotton); *perforatum; *hirsutum.
*Lamium Galeobdolon.
Limosella aquatica (ditch, Holmwood).
*Orchis mascula (woods).
*Oxalis Acetosella.
Populus tremula.
*Pyrus Aucuparia.
*Rosa systyla (woods).
*Rubus carpinifolius; *corylifolius (woods).
Ruscus aculeatus.
*Scabiosa succisa.
*Scilla nutans.
*Senebiera didyma (Dorking road).
Senecio sylvatica.
*Solidago Virgaurea.
*Stachys sylvatica.
*Tamus communis.
*Teucrium Scorodonia.
Tulipa sylvestris (orchard above).
*Vaccinium Myrtillus (Ridland Hill).
*Viburnum Opulus.
*Viola odorata (lanes, Leith Hill, and swampy hollows below).

LEITH HILL.

*Aira flexuosa; caryophyllacea; præcox.
*Anagallis tenella (bogs).
Campanula rapunculoides (towards Wotton).
*Carex glauca; *flava; binervis; *cæspitosa; pallescens (near Peslik).
Corydalis claviculata (moist woods about Abinger, &c.).
Digitalis purpurea (copses).
Eriophoron angustifolium; *vaginatum.
Gnaphalium sylvaticum.
*Hieracium umbellatum; murorum; boreale (banks towards Dorking).
Hypericum Androsæmum (near Lonesome); *Elodes (bogs).
*Juncus obtusiflorus; squarrosus.
Luzula multiflora (ravine).
*Molinia cærulea (moory parts).
Myosotis repens (bogs).
*Nardus stricta.
*Narthecium ossifragum (bogs).
*Orchis maculata (swampy places).
*Rhamnus Frangula.
Rubus suberectus; carpinifolius; villicaulis; tomentosa.
*Scutellaria minor.
*Solidago Virgaurea.
*Triodia decumbens.
*Ulex Gallii (with heaths, &c.).
*Vaccinium Myrtillus (abundant).
Veronica montana.

CRYPTOGAMS (GENERALLY).

Aspidium aculeatum (swampy wood N. of Cold Harbour, also var. lobatum).
Aulocomnion palustre and Sphagnum.
Botrychium Lunaria (summit).
Hypnum cuspidatum, &c.
Lomaria Spicant; Pteris.
Lycopodium clavatum (near the summit); *inundatum; Selago (wet places, foot of hill).
Mnium undulatum.
Nephrodium spinulosum; dilatatum (moist woods); Oreopteris (Holmwood, Leith Hill descent towards Wotton, and Broadmoor, foot of the hill).

Osmunda regalis (near Lonesome (?) olim; between Dorking and Cold Harbour; Peslik Bottom).
Pilularia globulifera (large pond on the Holmwood).
Splachnum ampullaceum (Leith Hill).
Trichostomum canescens.

59. TILGATE FOREST.

This extensive forest in the Weald of Sussex lies south of the branch line from Three Bridges *via* Crawley to Horsham. Oak and Birch, with plantations of Firs, is the prevailing timber. The damp hollows and sandstone ridges within its precincts produce several plants rarely seen nearer London. The forest may be reached by the Balcombe and Cuckfield road at Crawley, or from the Three Bridges station direct into this road over Pound Hill. Subsoil, calcareous sandstone and grit; in places, argillaceous.

Aquilegia vulgaris (copse right).
Carex vesicaria (copse right).
Epipactis latifolia; var. purpurata.
Erythræa pulchella (banks of ponds near Three Bridges station).
Cicendia filiformis (bog beyond Pease Pottage Gate).
Fragaria elatior.
Gentiana Pneumonanthe (several places near Handcross, and between the 'Norfolk Arms' and Balcombe).
Gymnadenia albida.
Hypericum Androsæmum (near Cuckfield, in the forest).
Luzula Forsteri; sylvestris.
Narcissus biflorus.
Narthecium ossifragum (in the hollows).
Serratula tinctoria.
Wahlenbergia hederacea, &c.

CRYPTOGAMS.

Asplenium Adiantum-nigrum.
Athyrium Filix-fœmina.
Hymenophyllum tunbridgense (on the main ridge near Balcombe, and on Chiddingly Rocks, Pook Church Rock).
Lomaria Spicant.
Lycopodium clavatum; inundatum; Selago (all on the banks of a pond below the bog between Pease Pottage Gate and Starvemouse Plain).
Nephrodium spinulosum; dilatatum; Oreopteris (banks of rills); Filix-mas.
Polypodium Phegopteris.
Scolopendrium vulgare.

60. FELBRIDGE.

"Twenty-five miles from London, between Godstone and East Grinstead," and "abounding with large pools of water, deep ravines, rocky chasms, woods, bogs, and heaths;" easily accessible from either of the above stations. Subsoil argillaceous and sandy, with calcareous grit in places.

Actinocarpus Damasonium (Hedge Pool).
Bupleurum rotundifolium (fields about).
Carduus pratensis.
Carex stellulata; pulicaris; panicea; lævigata; paniculata; Pseudo-cyperus; ampullacea.
Chlora perfoliata (fields).
Convallaria majalis (wood by Furnace Pool).
Cuscuta Epithymum.
Drosera rotundifolia; intermedia.
Erythræa Centaurium.
Genista anglica.
Lathyrus Nissolia (woods).
Littorella lacustris (Hedge Pool).
Lysimachia vulgaris (hedge ditches).
Melampyrum pratense (woods).
Melittis Melissophyllum (and varieties, wood adjoining Furnace Pool).
Menyanthes trifoliata (Woodcock Pool).
Myrica Gale (wood towards Crawley).
Myriophyllum spicatum (Woodcock Pool).
Narthecium ossifragum (bogs).
Neottia Nidus-avis.
Nymphæa alba (Woodcock Pool).
Orobanche major; minor.
Potamogeton.
Prunus domestica (near the mill, lower end of Woodcock Pool).
Pyrus Aucuparia (chasms of rocks); torminalis (woods).
Radiola Millegrana.

Rhamnus Frangula.
Ribes nigrum (chasms of rocks).
Scirpus lacustris.
Scutellaria galericulata.
Typha latifolia; angustifolia.
Utricularia (species not known).
Viola palustris (Furnace Pool).
Wahlenbergia hederacea (borders of Woodcock Pool and in a wet, boggy field).

61. HIGH ROCKS AND WATERDOWN FOREST, NEAR TUNBRIDGE WELLS.

An argillaceous soil, more or less mixed with calcareous grit and sandstone rocks in parallel ridges (as met with in Tilgate Forest and elsewhere in the Weald), characterises the country about Tunbridge Wells; damp hollows, rocky ravines, and occasionally patches of bog are frequent in the neighbouring woods and forests. Rare plants are to be found in this locality, especially Ferns, Mosses, and Scale-mosses.

Actinocarpus Damasonium (forest).
Agrostis setacea (on one spot in the forest, near Heathfield).
Aquilegia vulgaris (near High Rocks).
Calamagrostis Epigejos (wood near High Rocks, towards Tunbridge Wells).
Carex dioica (peat bogs; forest); lævigata (woods about).
Corydalis claviculata (High Rocks).
Delphinium Ajacis (by the stream).
Dentaria bulbifera (both sides of the stream at High Rocks).
Erythræa pulchella (cornfields between road to High Rocks and Rusthall Common).
Euphorbia platyphylla (road to the rocks).
Gastridium lendigerum (cornfields near).
Gentiana Pneumonanthe (sides of bogs in the forest towards Erith Park; also field right of coach road over the forest to High Rocks); campestris (in the forest).
Gnaphalium sylvaticum (High Rocks).
Hypericum Androsæmum (copses about).
Hypochœris glabra (Tunbridge Wells Common).
Inula Conyza (High Rocks).
Lithospermum officinale (High Rocks).
Malaxis paludosa (great bog near Kidbrooke Park gates).
Menyanthes trifoliata (bogs, forest).
Myrica Gale (in the forest).
Rhynchospora alba (bogs in the forest).
Sedum Telephium (banks of the stream, High Rocks).
Trifolium ochroleucum (banks of the stream, High Rocks).

CRYPTOGAMS.

Andræa rupestris (High Rocks).
Asplenium lanceolatum (rocks, and by path leading to High Rocks); Trichomanes (rock, Tunbridge Wells Common).
Diphyscum foliosum (sand-rocks, Eridge).
Encalypta vulgaris (fissures of the rocks).
Grimmia apocarpa (sand-rocks).
Hookeria lucens.
Hymenophyllum tunbridgense (High Rocks).
Splachnum ampullaceum (high part of Ashdown Forest, not far from Wych Cross, and in the great bog, Forest Row).
Trichostoma canescens (sandy, hilly places near Tunbridge Wells); heterostichum (High Rocks).

62. HILLS EAST OF WROTHAM.

At Wrotham the downs incline to the north-eastward, and thus form the western side of a funnel, leading to the outlet by which the river Medway finds its way to the estuary of the Thames. On the opposite side of this funnel, near Rochester, the hills appear to attain a greater elevation than elsewhere along the range, although in the immediate vicinity of Wrotham they are not less than eight hundred feet high. The same smooth rounded outline of configuration characterises this section of the downs as elsewhere; the same vales of drainage deepening and widening as they descend; the same sheepwalks above, sprinkled here and there with Juniper, and crowned with beechwoods; below, the wheat-producing line of the Gault, and beyond, the sandstone-ridges of the Wealden. A shady lane, grass-grown, and little used, leads along the foot of the downs from Wrotham onwards, whence there is easy access to the slopes and such of the woods and copses as are unenclosed. This lane serves, as a matter of fact, to form a boundary line between the formations, and as a limit to the plants peculiar to that of the chalk; which appear here side by side with those of the ordinary woodland type.

Alchemilla vulgaris (Hill Park near Westerham).
*Anthyllis Vulneraria.
*Asperula cynanchica.
*Brachypodium pinnatum (downs); *sylvaticum.
*Briza media.
*Bromus erectus; *racemosus; *asper.
*Calamintha menthifolia (plentiful); *Clinopodium.
*Campanula Trachelium; *glomerata.
*Carduus acaulis.
*Carlina vulgaris.
*Centaurea Scabiosa.
*Cephalanthera grandiflora.
*Chlora perfoliata (in profusion in some places).
*Cichorium Intybus.
*Cistus Helianthemum.
*Clematis Vitalba.
*Digitalis purpurea.
*Dipsacus pilosus (in the lane two miles beyond Wrotham; a patch of it); sylvestris.
*Echium vulgare.
*Erigeron acris.
*Erythræa Centaurium.
*Euphorbia platyphylla (cornfields); *amygdaloides (hedgebanks).
*Euphrasia officinalis.
*Galeopsis Tetrahit; *Ladanum.
*Gentiana Amarella.
*Helminthia echioides.
*Hypericum hirsutum; *perfoliatum.
*Inula Conyza.
*Iris fœtidissima.
*Juniperus communis.
*Linum catharticum.
*Melampyrum pratense (hedge below).
*Orchis pyramidalis.
*Origanum vulgare.
*Pastinaca sativa.
*Picris hieracioides.
*Poterium Sanguisorba.
*Primula vulgaris (hedges below).
*Reseda lutea.
*Rosa canina; *micrantha.
Salvia pratensis (near Wrotham in private grounds; Mr. Hanbury).
*Scabiosa arvensis; *succisa; *Columbaria.
*Senecio sylvaticus; *erucifolius.
*Tamus europæus.
*Taxus baccata.
*Teucrium Scorodonia.
*Thymus Serpyllum.
*Verbascum Thapsus.
*Viburnum Lantana.
*Viola odorata; *hirta; *sylvestris.

63. COBHAM AND CUXTON.

From Cobham to Cuxton the distance by the road is about four miles. The road lies in a hollow below Cobham Park, from which it is separated by a cultivated slope. A footpath, however, runs along the margin of these extensive grounds, with ready access to the road beyond. Towards Cuxton, from the hamlet of Bush onwards, the slopes are uncultivated and of low elevation, though steep. Subsoil, chalk and calcareous grit, producing the usual chalk plants. At Cuxton are mud-banks and marshy flats, with many plants peculiar to the shores of a tidal river.

Adonis autumnalis?
*Ajuga Chamæpitys (fields).
*Asperula cynanchica.
*Astragalus Glycyphyllos (copses, park borders).
*Brachypodium pinnatum (in the park).
*Briza media (in the park).
*Calamintha Clinopodium; *menthifolia (roadside beyond Bush).
Cephalanthera grandiflora; ensifolia.
*Chlora perfoliata.
*Cichorium Intybus.

*Clematis Vitalba.
Dianthus Armeria.
Euphorbia Lathyris (near Cobham).
*Galeopsis Ladanum (fields).
*Helianthemum vulgare.
*Hippocrepis comosa.
Lathyrus Nissolia.
*Linaria spuria.
Lithospermum arvense.
Monotropa Hypopitys (woods).
*Nepeta Cataria (at Bush).
Ophrys apifera; muscifera.
Orchis pyramidalis; fusca.
*Origanum vulgare.
*Papaver somniferum (field).
*Pastinaca sativa.
*Petroselinum segetum (in field sloping up to the park, abundant).
*Picris hieracioides.
*Reseda lutea.
Rhamnus catharticus.
Rosa spinosissima; rubiginosa; tomentosa.
*Scabiosa arvensis.
*Sclerochloa rigida.
Verbascum Lychnitis; Blattaria?
*Verbena officinalis, &c.
*Viburnum Lantana.
Viscum album (on Thorn-trees).

Note.—Althæa hirsuta and Salvia pratensis are no longer found in or about the park.

AT CUXTON.

*Agrostis alba (stolonifera).
*Apium graveolens (ditches near).
*Aster Tripolium (abundant).
*Atriplex erecta.
*Ruppia maritima (ditches near).
*Scirpus maritimus.
*Sclerochloa maritima.
*Spergularia marina (neglecta).
Near Cuxton, Stroud Road: Linum angustifolium.
Chalk-pits near Rochester: Verbascum Blattaria.
*Melilotus alba.
*Triglochin maritima.

ADDENDA.

Castle-walls, Rochester: *Dianthus caryophyllus. Flats beyond: Salicornia; Beta maritimâ, Statice, &c. Marshes, Stroud: Bupleurum tenuissimum.

64. ABOUT NORTHFLEET AND GRAVESEND.

Northfleet and Gravesend are in juxtaposition, so much so that the houses are continuous above the chalk banks which form the river frontage. The excavations are of immense extent, and about Gravesend especially are replete with verdure and foliage. Beyond this place in the direction of Cobham, is gently rising ground, with a subsoil of chalky grit and cultivated; but by the river-side wide flats are seen, now drained and converted into pasturage (formerly marshy), extending beyond Higham and Cliffe to the mouth of the Medway; near Northfleet are also flats by the river, and at Swanscombe is a wood rich in plants, as are also the roads and lanes inland through Southfleet and Betsome. This locality, therefore, is a good one for field botany, and of easy access.

Aceras anthropophora.
Agrostis Spica-Venti (Northfleet).
*Ajuga reptans (Higham pit, profuse).
Alopecurus bulbosus (Northfleet).
*Apium graveolens.
*Artemisia maritima; Absinthium (near pits?).
*Asparagus officinalis.
*Asperula cynanchica (cliffs).
Aster Tripolium.
*Atriplex arenaria; littoralis; portulacoides.
Avena pubescens (chalky hills).
*Beta maritima.
*Campanula glomerata (chalky banks).
*Carex divisa (flats); *flava (pit bottoms).
Catabrosa aquatica (ditches).
Caucalis daucoides.
*Centaurea Calcitrapa (roadside, &c., near the cement works).
*Centranthus ruber (pits; also a white variety).
Cheiranthus Cheiri (cliffs).
Chenopodium hybridum (about N'fleet).

*Chlora perfoliata (pits).
*Cichorium Intybus.
*Clematis Vitalba.
*Crepis taraxacifolia; fœtida; *biennis.
*Cynoglossum officinale (abundant by the canal 1¼ mile from Gravesend).
*Cynosurus cristatus (pit bottoms).
Dianthus Armeria (between Shorne and Stroud).
Digitalis purpurea (woods and copses).
*Diplotaxis tenuifolia (pits, abundant).
*Echium vulgare.
Epipactis latifolia (woods at Cliffe).
*Erythræa Centaurium (also at Shorne, &c.).
*Euphorbia platyphylla (fields).
*Festuca sciuroides (flats, Northfleet).
Fœniculum vulgare.
Galium tricorne (fields towards Cobham); anglicum (chalk-cliffs).
Gastridium lendigerum (cliffs).
Gentiana Amarella (pits).
Helleborus fœtidus (chalk-cliffs).
*Helminthia echioides (Higham, &c.).
Hieracium sylvaticum.
*Hippocrepis comosa (on the chalk).
*Hordeum maritimum; *pratense.
*Hydrocharis Morsus-ranæ (ditches, flats at Higham, abundant).
*Hypericum hirsutum; *perforatum.
*Inula Conyza (pits, &c.).
Iris fœtidissima (woods).
*Juncus maritimus.
Lactuca saligna (near Cliffe).
*Lemna polyrhiza (marsh ditches).
*Lepidium ruderale.
*Lepturus filiformis (plentiful 1½ miles or so from Gravesend by the canal).
*Linaria minor (pits); *spuria (fields); repens (cliffs).
*Linum catharticum (pits).
Listera ovata.
Lithospermum officinale; *arvense.
*Lysimachia Nummularia (pit bottoms).
*Melilotus officinalis.
Myosurus minimus (cliffs).
*Nepeta Cataria (roadside, Northfleet).
*Ononis arvensis.
Onopordum Acanthium.
*Ophrys apifera; muscifera; aranifera?
Orchis pyramidalis; fusca; *maculata.
*Papaver hybridum (fields towards Cobham).
*Parietaria diffusa (cliffs, abundant).
*Pastinaca sativa.
*Petroselinum segetum (cliffs).
*Pimpinella Saxifraga.
*Potamogeton pusillus (ditches, flats).
*Poterium Sanguisorba.
*Reseda lutea; *Luteola.
*Samolus Valerandi (ditches on the flats).
*Saponaria officinalis (lane leading from Shorne to the flats, plenty).
*Scabiosa Columbaria.
*Scandix Pecten-Veneris (fields).
Scilla autumnalis (Shorne Warren).
*Scirpus maritimus.
Sclerochloa rigida; *maritima.
*Sedum sexangulare (old walls, Higham); *reflexum.
*Sison Amomum (Shorne).
Sium latifolium (Northfleet).
*Smyrnium Olusatrum (pits, Northfleet).
*Sparganium ramosum (ditches on the flats).
Specularia hybrida (fields).
*Spergularia neglecta.
*Statice Limonium (by the Thames).
*Torilis nodosa (flats, and on banks).
Trifolium maritimum; scabrum; striatum.
*Triglochin maritimum; *palustre.
*Triticum junceum (flats).
*Verbascum Thapsus.
*Viburnum Lantana.
*Zannichellia palustris (ditches on the flats).

CRYPTOGAMS.

Anomodon viticulosum.
Ceterach officinarum (Cliffe Church).
Homalothecium sericeum.
Hypnum lutescens; filicinum; Crista-Castrensis (? molluscum); nitens.
Neckera crispa (chalk cliffs, Gravesend).
Thuidium abietinum.

65. GRAYS AND TILBURY.

These localities opposite Northfleet and Gravesend, on the Essex side, are somewhat similar as regards the flats by the river and bordering chalk. The country however, comparatively speaking, is a low-lying one, for although higher ground is apparent at no great distance inland, it can hardly be regarded as hilly. The chalk banks, with extensive recent excavations at Grays, are little above the level of the plain, and are covered with two or three feet of gravel drift. Bordering the flats about Grays are cornfields; about Tilbury Fort these flats have somewhat of a fenny character, but less marshy now perhaps than was formerly the case, owing to drainage and the embankment here as elsewhere by the Thames.

*Aster Tripolium.
*Atriplex littoralis; *portulacoides.
*Bupleurum tenuissimum.
*Carex divisa.
*Centaurea Calcitrapa.
Chenopodium rubrum.
*Cynoglossum montanum (near Tilbury).
*Daucus Carota.
Festuca sciuroides (E. Tilbury); Pseudo-Myurus (ditto).
*Hordeum pratense; *maritimum.
*Juncus maritimus.
Lepidium latifolium? (Grays).
*Lepturus filiformis.
*Petroselinum segetum.
*Phragmites communis (ditches).
*Plantago maritima.
Polypogon monspeliensis? (opposite Northfleet, olim).
*Salicornia herbacea.
Sambucus Ebulus (between E. and W. Tilbury).
Schœnus nigricans? (fenny flats about the Fort, olim).
*Scirpus maritimus.
Sclerochloa procumbens; *maritima.
*Spergularia neglecta.
*Statice Limonium.
*Suæda maritima.
*Tordylium maximum (E. de C.).
Trifolium maritimum.
*Triglochin maritimum.
*Triticum junceum.

66. SOUTHEND AND CANVEY ISLAND.

Southend is beyond the limits of a thirty or thirty-five miles radius from the metropolis, but the facility of access to the place is so great, and the opportunity of obtaining many plants found only on the sea-shore so favourable, that the intrusion of the locality may be freely excused. High clay banks between Leigh and Southend, with muddy shore and a stone embankment by the railway; near Leigh a miniature sandy bay. At Canvey Island an alluvial flat, with muddy shores and for the most part cultivated; beyond Southend, towards Shoeburyness sandbanks, a muddy shore on one side and an alluvial flat on the other: such are the topographical characters of the localities. The flora is varied; and the occurrence of chalk plants indicates the presence of that substance in the subsoil.

Allium vineale.
Althæa officinalis.
*Armeria maritima (near Leigh).
*Artemisia maritima.
*Aster Tripolium.
Atriplex littoralis; *portulacoides; *Babingtonii; *arenaria.
*Beta maritima.
Bromus arvensis (Southend).
Cakile maritima.
*Carex divisa; arenaria.
*Carlina vulgaris.
Cerastium tetrandrum.
Chenopodium olidum; rubrum.
*Cochlearia officinalis.
Convolvulus Soldanella.
Coriandrum sativum.
Crepis biennis; taraxacifolia.
Dianthus Armeria (near Leigh, among bushes on the low hills).
*Diplotaxis tenuifolia; muralis.
Elymus arenarius.
Erigeron acris.
*Eryngium maritimum.
*Erythræa Centaurium (high clay bank); pulchella (near Southend).
Euphorbia platyphylla (cornfields above Southend); Paralias (S. Shoebury common).
Festuca uniglumis (Southend).
*Fœniculum vulgare.
Frankenia lævis (shore near Wakering).
*Geranium Columbinum (Southend).
*Glaucium luteum.
*Glaux maritima.
Honckeneya peploides (N. Shoebury).
*Hordeum maritimum.
Hyoscyamus niger.

Inula Conyza; crithmoides (salt marshes).
Kœleria cristata (commons near).
*Lactuca saligna; *virosa; Scariola.
*Lathyrus Nissolia (clay cliffs); tuberosus (Canvey Island); hirsutus (bushes about Hadleigh Castle).
Lavatera arborea (clay cliffs).
*Lepidium ruderale (Canvey Island).
*Linaria spuria (cornfields).
*Linum angustifolium (clay cliffs).
Lithospermum officinale.
Medicago minima (shore below Southend).
*Ononis spinosa.
*Onopordum Acanthium (shore below Southend).
Phleum arenarium.
*Phragmites communis.
*Picris hieracioides.
*Polypogon littoralis (Canvey Island); monspeliensis (Canvey Island).
*Prunus spinosa.
Psamma arenaria (Shoebury beach).
Ranunculus hirsutus (Canvey Island).
Reseda lutea.
Rubus corylifolius.
*Ruppia rostellata (ditches in the flats).
*Sagina maritima.
*Salicornia herbacea.
*Salix cinerea.
*Salsola Kali.
Samolus Valerandi.
Saxifraga granulata.
*Scirpus maritimus.
*Senecio erucifolius; sylvaticus.
Silene maritima.
*Sinapis nigra (clay cliffs).
Specularia hybrida (cornfields).
*Spergularia neglecta.
Stachys arvensis (cliffs towards Leigh).
*Statice Limonium.
*Suæda maritima.
Thymus Serpyllum.
Trifolium scabrum.
*Triglochin maritimum.
*Triticum junceum.
*Typha angustifolia (ditches in Canvey Island).
Verbascum nigrum.
Viola hirta.
Zostera marina.

67. NORTON HEATH, ONGAR AND FYFIELD.

Whatever may have been the extent of this heath formerly, but little remains of it at present but a few square acres of furze-grown common, left of the Chelmsford road. The subsoil is gravel, and many pits are upon the heath from which gravel has been dug out. Ongar and Fyfield are in the near neighbourhood, distant about four miles from each other and from the heath, in such wise that Fyfield forms the apex of the triangular space which they enclose. The country lies high hereabouts, and is well cultivated; it is drained by the upper waters of the Roding. The occurrence towards Fyfield of some plants belonging to the chalk series is accounted for by the presence of calcareous grit in the loamy gravelly soil.

*Achillea Ptarmica.
*Adoxa moschatellina (about Fyfield).
Alchemilla vulgaris (Great Canfield, near High Roding).
Allium vineale.
*Anthemis nobilis (on the heath).
Anthriscus vulgaris (about Fyfield).
*Betonica officinalis (Norton Heath).
*Bupleurum rotundifolium (cornfields, Fyfield); *falcatum (roadside beyond Norton Heath, towards Chelmsford).
*Calluna vulgaris.
Campanula glomerata (Fyfield).
Carex pendula (Norton Heath).
Carlina vulgaris (Fyfield).
Cerastium aquaticum (ditches, Ongar).
Chlora perfoliata (Fyfield).
Cuscuta Trifolii (Fyfield).
*Cynoglossum montanum (Fyfield).
Doronicum plantagineum (Fyfield).
Euphorbia platyphylla (cornfields, Fyfield).
Galanthus nivalis (field near C. Ongar).
Galeopsis Ladanum (cornfields).
Galium tricorne (cornfields, Fyfield).
Gentiana Amarella (Fyfield).
Gymnadenia conopsea (between High and Chipping Ongar).
Habenaria chlorantha; viridis (Norton Heath).
*Hydrocotyle vulgaris (Norton Heath).
Hyoscyamus niger (Fyfield).
Juncus obtusiflorus, &c. (C. Ongar).
Lathyrus Aphaca (near Ongar); Nissolia (Norton Heath); tuberosus (hedges and cornfields, Fyfield).

Lithospermum officinale (Fyfield).
*Lotus corniculatus; *major (Norton Heath).
Lysimachia vulgaris (by the Roding, Fyfield).
*Mentha sativa (Norton Heath).
*Myriophyllum verticillatum (Norton Heath).
Onopordum Acanthium (Fyfield).
Orchis latifolia (Fyfield).
Parnassia palustris (Chipping Ongar, and between Chipping and High Ongar by the Roding in marshy meadows).
*Picris hieracioides (towards Fyfield).
Pimpinella magna (Ongar).
*Potentilla anserina, &c. (Norton Heath).
Ribes Grossularia (Ongar).
Rumex pulcher (High Ongar).
Sagittaria sagittifolia (Ongar).
*Senecio sylvaticus (Norton Heath); *erucifolius (Fyfield).
Silybum Marianum (Fyfield).
Stachys arvensis (cornfields, Fyfield).
Vinca minor (Fyfield).
*Zannichellia palustris (Norton Heath).

CRYPTOGAMS.

Asplenium Adiantum-nigrum (?)(Norton Heath); Trichomanes (near Ongar).
Athyrium Filix-fœmina (Fyfield).
Ophioglossum vulgatum (between High and Chipping Ongar).

68. ESSEX CORNFIELDS.

A rolling expanse of undulating country, mostly cornfields, extends from beyond Fyfield in every direction. The section north-westward between that place and Bishop's Stortford will afford opportunities of observing what may be considered the characteristic flora. The subsoil is a gravelly loam, more or less mixed with calcareous grit; paths through the cornfields and lanes everywhere, from hamlet to hamlet, render access easy. Near the village of Hatfield Broad Oak is an extensive forest, private property, and not open to the public without permission.

*Æthusa Cynapium.
*Alopecurus agrestis.
*Anchusa arvensis.
*Avena fatua (cornfields).
*Bartsia Odontites (lanes and borders of cornfields).
Blysmus compressus (Hatfield Forest).
*Brachypodium pinnatum (park).
*Briza media (park, Hatfield Forest).
*Bromus secalinus (in two or three fields, Little Laver).
*Bupleurum rotundifolium (cornfields, frequent).
*Calamintha Clinopodium (lanes).
*Campanula Trachelium (roadside hedges near Hatfield Broad Oak).
*Carduus acaulis (in the park).
*Carex pendula (roadside, ditches); Pseudo-cyperus (lake in the park).
*Centaurea Cyanus.
*Cichorium Intybus (lanes).
*Clematis Vitalba (lanes).
*Convolvulus arvensis (abundant).
*Cynoglossum montanum (Hatfield Forest).
*Daucus Carota (roadsides).
Euphorbia platyphylla; *exigua.
*Fumaria officinalis (fields).
*Galeopsis Ladanum; *Tetrahit.
*Galium tricorne (fields).
*Helianthemum vulgare (roadsides).
*Lithospermum arvense (fields).
*Lythrum Salicaria (ditches, roadsides).
*Melilotus officinalis (roadsides).
*Ononis arvensis (roadsides).
*Papaver Rhœas (fields).
*Picris hieracioides (roadsides).
*Ranunculus arvensis (fields).
*Sagina nodosa (by the lake in the park).
*Scabiosa succisa; *arvensis.
*Trifolium fragiferum (roadsides).
*Viburnum Lantana (hedges).

69. HERTFORD HEATH AND SURROUNDING WOODS.

Hertford Heath is situated on high ground, between Hoddesdon and Hertford. It is surrounded by woods, and has nothing in its aspect in common with the sandy heaths south of the Thames, but, from being much overgrown with bushes, has a great resemblance to Epping Forest. It is, however, of a more open character to the left of the Hertford road, and this portion is separated from that which is right of the road by enclosures and the village. The scrub consists principally of Oak, Bramble, Briar, Blackthorn, and Whitethorn, Calluna and Salix repens. Subsoil gravel. Pryor's Wood and the woods bordering it in the direction of Haileybury College are principally of Oak and Hornbeam, with an admixture of Ash, Maple, and Sallow. Ball's Wood, near Hertford, is of considerable extent, and enclosed; it is a cold damp wood, sloping to the north; with a dense vegetation of Oak, Maple, Hornbeam, and Ash, and an undergrowth of Hazel, Sallows, Viburnum Opulus, Briar and Bramble. The gravelly subsoil of this part of Herts rests upon chalk. In the woods are patches of swamp and pools.

*Aira cæspitosa.
Allium vineale (between Hoddesdon and Haileybury).
Arabis hirsuta (copse, hillside beyond Goldings).
*Betonica officinalis.
*Brachypodium sylvaticum.
*Bromus asper; *racemosus (bordering lanes in profusion).
*Calamagrostis Epigejos (Ball's Wood and Pryor's Wood?); lanceolata? (Ball's Wood).
*Carex pendula; *sylvatica; *flava; *vesicaria (Pryor's Wood); binervis; pilulifera; strigosa (Quick's Hill Wood); Bœnninghausiana (ponds in Ball's Wood).
Chrysosplenium oppositifolium (Quick's Hill Wood).
Comarum palustre (Ball's Wood).
Convallaria majalis (Pryor's Wood).
*Dipsacus pilosus (ditches bordering hedges of Ball's Wood).
*Erythræa Centaurium (Ball's Wood).
*Genista tinctoria (borders of Ball's Wood).
Habenaria chlorantha (Ball's Wood and Box Wood).
*Helosciadium inundatum.
*Hieracium vulgatum; boreale (?) (Hertford Heath).
*Holcus lanatus.
*Hypericum hirsutum (Ball's Wood).
Iberis amara (field N. of Pryor's Wood.
Lysimachia Nummularia (Ball's Wood).
*Melampyrum pratense (woods).
Mentha sativa.
*Menyanthes trifoliata (Pryor's Wood). also in Ball's Wood).
*Milium effusum (woods).
Mœnchia erecta (Hertford Heath).
*Neottia Nidus-avis (woods).
Œnanthe Phellandrium (ponds N. side of Ball's Wood).
Orobanche major (Hertford Heath).
Paris quadrifolia (Ball's Wood).
Pimpinella magna (about Ball's Wood, and towards Bayford).
*Poa nemoralis (woods).
*Populus tremula (woods).
*Rosa arvensis.
*Rubus glandulosus.
*Serratula tinctoria (borders of Ball's Wood).
Stellaria glauca (Hertford Heath).
Vinca minor (Box Wood).

CRYPTOGAMS.

Asplenium Adiantum-nigrum (Hertford Heath, near Townsend Arms).
Athyrium Filix-fœmina (?) (Pryor's Wood).
Lomaria Spicant (Hertford Heath).

70. THE LEA VALLEY ABOUT HATFIELD, HERTFORD AND WARE.

The river Lea, from Hatfield onwards, pursues an easterly course towards Ware; meandering through meadows with slopes on the north side more generally cultivated, while those to the southward are principally wooded. Ball's Wood (referred to in the preceding locality), the woods of Bayfordbury, Brickendonbury, about Essendon and at Hatfield Park, are all private property and enclosed, and unless where crossed by a public road or footpath, cannot be entered without permission. North of the river are Ware and Panshanger Parks. The Lea is an inconsiderable stream to within a short distance of Hertford, although swelled by the drainage from the uplands to the south of it; near Hertford it is joined by two affluents. The subsoil is alluvial in the valley; gravelly with underlying chalk on the slopes. Near Hertford and east of Hatfield Park are chalk pits; these, with the woods, where accessible, and the waste gravelly fields of occasional occurrence, the roadsides, and meadows by the river, offer opportunities of obtaining specimens of a few plants not frequent elsewhere near London.

*Adoxa moschatellina (woods).
Alchemilla vulgaris (near Essendon and Brickendon, and meadows near Stanborough).
Allium vincale (dry knolls in the meadows between Hertford and Ware).
Anthemis arvensis (Gallows Plain, Hertford).
*Anthyllis Vulneraria (chalk pits Hertford, Essendon, and near Hatfield Park).
Arabis perfoliata (frequent N.W. of Hertford, and between Hatfield Park and Cole Green).
Asperula cynanchica (pits, Hertford).
Atropa Belladonna (Hatfield Park).
Barbarea arcuata (between Hertford and Ware Park).
*Brachypodium pinnatum (chalk pit near Hertford).
Bupleurum rotundifolium (Mangrove Lane, fields adjoining the brook).
Calamintha Acinos (Gallows Plain).
Campanula glomerata (pit behind Hertford Union workhouse).
Carduus acaulis (about Hertford, scarce); *nutans (about Hertford, scarce).
Carex acuta; Pseudo-cyperus (between Hertford and Ware); pallescens (ponds S. and S.W. of Hertford); pulicaris (woods S. of Hertford); strigosa (Stanborough and elsewhere).
Carlina vulgaris (about Hertford, scarce).
Centaurea Cyanus (fields, Hertford, scarce).
Ceratophyllum aquaticum (between Hertford and Ware).
Chlora perfoliata (pit near Hatfield Park).
*Clematis Vitalba (about the chalk pits).
Cynoglossum officinale (Hatfield Park and Ware Park).
Dianthus Armeria (Mangrove Lane; Gallows Plain; gravel pits near Hatfield).
Dipsacus pilosus (about Bayford, Essendon and Little Berkhampstead).
Drosera rotundifolia (Hatfield woodside).
Eleocharis multicaulis (bog by Kentish Lane, Hatfield).
Epilobium roseum (Bayford, near the church, in a ditch by the road to Hertford).
Epipactis latifolia (woods, south); palustris (wet pasture by the brook, East-end Green, Hertingfordbury).
Erigeron acris (gravel pit, Welwyn road and about Bayford, &c.).
Festuca sciuroides (between Hertford and Ware).
Filago gallica (field right of the road from Hertford to Welwyn, and footpath between Bayford House and Bayford wood).
Fumaria capreolata (field near Easney Park wood; footpath to Ware).
Galium tricorne (fields near).
Gastridium lendigerum (gravelly field near N.E. boundary of Hatfield Park).
Genista tinctoria (about Bayford and Little Berkhampstead).
Geranium columbinum (about Hertford).
Helianthemum vulgare (pits, Essendonbury and near Hatfield Park, Gallows Plain).
Helleborus viridis (Watery Hall Farm, near Hertingfordbury, in woods on the chalk).
Herminium Monorchis (chalk-pit E. side of Hatfield Park).
Hieracium boreale (?) (frequent, near Hertford, Thieves' Lane, Hertingfordbury).
Hottonia palustris (between Hertford and Ware).
Hyoscyamus niger (Mead Lane, Hertford; Ware Park; Hatfield Park).
Hypericum Androsæmum (lanes about Essendon, Little Berkhampstead and Bayford).
Inula Helenium (Mangrove Lane, by a pond near Blue Close).
Juncus diffusus (near Cole Green).
Kœleria cristata (between Hertford and Ware).
Mentha rotundifolia (between Hertford and Essendon, opposite Watery Hall Farm).
*Nepeta Cataria (by pit near Hertford).
Nymphæa alba (in the Lea at Bayfordbury Farm).
*Œnanthe fluviatilis (in the Lea).
Onopordum Acanthium (Ware road, in a field between Gallows Plain and Ball's Park).
Ophrys apifera (field towards Bayford, one mile from Hertford).

Orchis Morio (pastures); pyramidalis (chalk pits).
Ornithogalum umbellatum (meadows by the Lea towards Ware).
Orobanche elatior (steep banks between the Ware road and Gallows Plain).
Papaver Argemone (fields N.W. of Hertford, also on Gallows Plain and about Essendon).
*Parnassia palustris (Lea valley, several places; Stanborough; near Hatfield; and between Cole Green and Hertingfordbury).
Phleum Bœhmeri (near Hertford Union Workhouse, on a steep gravelly bank, road to Stanstead).
Petroselinum segetum (between Hertford and Hertingfordbury).
*Plantago media (chalk pits).
Potamogeton lucens; crispus; pusillus; pectinatus; densus (in the Lea).
*Poterium Sanguisorba (chalk pits).
Pyrus Aria (Hatfield Park); torminalis (Bayford).
Ranunculus parviflorus (Essendon, near the church and elsewhere about).
Ruscus aculeatus (Milwards Park Wood, Hatfield: woods, Essendon).
Salix pentandra (riverside near Whitwell).
Saponaria officinalis (hedge near Roxford Farm, Hertingfordbury).
Saxifraga granulata (Mead Lane, Hertford, and meadows between Hertford and Ware; road to Stanstead).
Scabiosa arvensis; Columbaria (about Hertford).
Scandix Pecten-Veneris (fields).
Scrophularia vernalis (in Hatfield Park; also hedge near the gasworks, Hatfield)?
Sedum reflexum (Hertford Castle); Telephium (near Hertford, Bayford, Essendon, &c.).
Silybum Marianum (Ware road, field between Gallows Plain and Ball's Park).
Sisymbrium Irio (near the gasworks, Hertford).
Verbascum nigrum (about Hertford).
*Viburnum Lantana (about the pits, &c.).
Vinca minor (Hertingfordbury Park and in Mole Wood).
Viola hirta (about the pits, &c.).

71. UPPER COLNE DISTRICT.

About Rickmansworth, Watford and St. Albans. Low chalk hills, capped with gravel drift, impart to this section of the Colne districts an undulating character. The river swollen by the Chess rivulet, and another affluent from the Berkhampstead vale, which joins it at Rickmansworth, is so far a considerable stream; beyond Watford it is comparatively insignificant. Meadows, often wet and moory, border it on either hand, and produce one or two plants of rare occurrence. The hillsides, in places where denuded of subsoil, are productive of chalk plants, which increase in frequency further northwards in the direction of St. Albans; but there are no elevated downs similar to those of Surrey and Kent, nearer than Tring, and the borders of Bedfordshire and Cambridgeshire. Only the more ordinary plants of this formation are, therefore, to be met with in the locality.

In addition to many of the plants enumerated in section 32, the following occur:—

Rickmansworth.

Alchemilla vulgaris (common moor).
Arabis perfoliata (Long Valley Wood).
Asplenium Trichomanes (old wall between Moor Park and the Colne).
Atropa Belladonna (woods).
Chrysosplenium oppositifolium (meadows below Coney Farm).
Cynoglossum officinale (pasture behind the Swan Inn).
Dentaria bulbifera (Loudwater Wood, and wood near High Wood).
Dipsacus pilosus (lane near Moor Hall).
Epilobium roseum (footpath to Scott's Bridge).
Epipactis latifolia.
Genista tinctoria (common moor).
Helleborus viridis (lane S.W.).
Hieracium murorum (woods).
Hordeum sylvaticum (Hill Wood).
Lithospermum officinale (Long Valley Wood).
Œnanthe crocata.
Ophioglossum vulgatum (meadows).

Ophrys apifera (woods).
Parnassia palustris (moors).
Polygonum Bistorta.
Saxifraga granulata (common moor and in Moor Park).
Stachys arvensis (fields between Long Valley Wood and the Watford road).

Watford, &c.

Alchemilla vulgaris (towards Bushey).
Allium vineale (meadows, Bourne End).
Anemone ranunculoides (Abbot's Langley).
Artemisia Absinthium (St. Albans road).
Calamagrostis Epigejos (copse over the S. mouth of the railway tunnel).
Cephalanthera grandiflora (wood S. of Bourne End).
Chenopodium Bonus-Henricus.
Chrysosplenium oppositifolium (opposite the Spring Bourne End).
Dianthus Armeria (railway banks north of the tunnel).
Epipactis palustris (roadside near Hemel Hempstead).
Galanthus nivalis (meadows, Bourne End Mill).
Geranium lucidum (lanes between the river and the Rickmansworth road).
Helleborus viridis (woods near the goods station).
Hippuris vulgaris (in the river).
Hordeum sylvaticum (Long Spring).
Hottonia palustris.
Hyoscyamus niger (Box Moor).
Medicago falcata (between Watford and Bushey Hill ? olim).
Menyanthes trifoliata (Box Moor).
Ophioglossum vulgatum (meadows).
Orchis latifolia (Box Moor).
Paris quadrifolia (Bourne End).
Parnassia palustris (meadows, Watford, near the railway arch, and beyond Bourne End).
Potamogeton perfoliatus.
Salix pentandra (roadside near King's Langley).
Saxifraga granulata (meadows beyond Bourne End).
Serratula tinctoria (Berry Wood).
Spiranthes autumnalis (Box Moor).
Stellaria aquatica (meadows).
Trifolium subterraneum (Box Moor).
Viburnum Lantana.
Vinca minor (lanes about King's Langley).

St. Albans.

Arabis perfoliata (Hatfield road).
Carex paniculata (Pondyards and Little Mill).
Carlina vulgaris.
Cheiranthus Cheiri (Abbey walls ? olim).
Diplotaxis tenuifolia (Abbey walls).
Erica cinerea (Præ Wood).
Gastridium lendigerum (waste field midway towards Hatfield).
Genista tinctoria (field below Verulam Buildings, Redbourne road).
Geranium pratense (meadows near Holywell Bridge); columbinum (St. Stephen's Hill).
Gnaphalium sylvaticum (woods).
Helianthemum vulgare (ruins of Verulam).
Hieracium murorum (Abbey walls); boreale (S.W. corner of Verulam).
Hypericum Androsæmum (Præ Wood).
Inula Helenium (between St. Albans and Hatfield).
Iris fœtidissima (roadside hedges, towards Dunstable).
Linaria repens (walls, Dagnall Lane).
Lithospermum officinale; arvensis (cornfields, especially near Oster Hills).
Orchis Morio (pastures).
Pyrus communis; torminalis.
Ranunculus parviflorus (cornfields).
Scabiosa Columbaria (Sandridge road).
Sedum Telephium (hedges).
Verbascum nigrum (about the ruins of Verulam, &c.)

72. TRING AND ALDBURY.

Rounded chalk hills, with or without beechwoods on their slopes and summits, characterise this locality, and in so far it resembles many sections of the Kentish and Surrey Downs. The flora is similar in character, with the addition, however, of two or three specialities. A somewhat enclosed valley, along which runs in a north-westerly direction the canal and the river, or rather rivulet, as well as the high road and the railway from Watford, past Berkhampstead, opens out at Tring station into a plain. The range, on the left tending south-westwards, on the right bends beyond Aldbury Newers Hill abruptly to the north and eventually north-east.

The village of Aldbury is in a hollow of the range right, so that Aldbury Nowers Hill stands out as a sort of promontory or headland, six or seven hundred feet above the plain. The town of Tring, two miles from the station in the opposite direction, lies at the foot of thickly-wooded slopes; near it, in the plain, are some large reservoirs, of which the borders are in part marshy. The flora is varied and the locality still good, but, owing to the rapacity of collectors, no longer so prolific as formerly. Subsoil, chalk, or gravelly.

Aceras anthropophora (chalk slopes beyond Tring).
*Adoxa moschatellina (woods).
Alchemilla vulgaris (heath south of Tring, olim; also in the woods N.W. of the monument, Aldbury. — N.B. Heath enclosed and ploughed up).
*Anemone Pulsatilla (downs, Aldbury Nowers).
Anthyllis Vulneraria.
*Aquilegia vulgaris (woods west).
*Asperula odorata; *cynanchica.
Atropa Belladonna (woods).
*Brachypodium pinnatum (downs).
*Bromus erectus; *racemosus.
*Bunium flexuosum (woods).
*Calamintha Clinopodium; *Acinos.
*Campanula glomerata; *Trachelium.
*Carduus nutans; *acaulis.
*Carex sylvatica.
*Carlina vulgaris.
*Cephalanthera grandiflora.
Chlora perfoliata.
*Circæa lutetiana.
*Clematis Vitalba.
Convallaria multiflora (bottom of Hanging wood, Tring, a little to right of path down the hill).
*Daphne Laureola.
*Epilobium angustifolium (Aldbury Nowers wood).
Epipactis latifolia.
Erythræa Centaurium.
*Euphorbia amygdaloides.
*Galeopsis Ladanum (cornfields); Tetrahit.
Gentiana Amarella.
Gnaphalium sylvaticum.
Gymnadenia conopsea.
Habenaria viridis (downs); bifolia.
*Helianthemum vulgare.
Helleborus viridis (between Tring and the reservoirs).
Hieracium umbellatum; vulgatum; boreale.
Hippocrepis comosa.
Hippuris vulgaris (reservoir).
*Hordeum sylvaticum.
*Hypericum hirsutum; *perforatum.
*Inula Conyza.
*Juniperus communis (Aldbury downs).
*Ilex aquifolium (woods).
Kœleria cristata (Tring).
*Lactuca muralis (in the woods).
*Lamium Galeobdolon.
*Linum catharticum.
Lithospermum officinale (hedge at Wilstone near Tring).
Melampyrum cristatum (wood near North Church common, Tring).
*Mercurialis perennis (in abundance in the woods).
*Milium effusum.
*Monotropa Hypopitys.
Myriophyllum verticillatum (reservoir).
*Neottia Nidus-avis.
Ophrys apifera; muscifera (Tring and Aldbury Nowers).
*Orchis pyramidalis; militaris; ustulata (Tring and Aldbury Nowers).
*Origanum vulgare.
Paris quadrifolia (woods, Aldbury).
Parnassia palustris (by the reservoir).
Pimpinella magna (hedge near Aldbury).
*Plantago media.
*Poterium Sanguisorba.
*Primula vulgaris; *veris.
*Prunus Cerasus (woods).
Pyrola minor (Tring woods and Aldbury Nowers wood, and copse by Shiere Lane; media (?) (beech woods about).
Pyrus Aria (woods about).
*Rubus carpinifolius; *fusco-ater; *Kœhleri; *macrophyllus; rudis; tomentosa; plicatus (corner of heath (?) olim); *Idæus (Aldbury Nowers wood); nitidus; leucostachys.
*Sanicula europæa.
*Scabiosa Columbaria.
*Sedum Telephium.
*Senecio campestris (downs right of the station).
Spiræa Filipendula.
*Thymus Serpyllum.
*Viburnum Lantana.
Vicia sylvatica.
Vinca minor (Wix's wood).
*Viola hirta.

CRYPTOGAMS.

Athyrium Filix-fœmina.
Hypnum molluscum, &c.
Lomaria Spicant (?) (on the heath, olim).
Nephrodium Filix-mas; Oreopteris (Shiere Lane).

ADDENDA.

Berkhampstead Common, Verbascum virgatum (?); *Digitalis purpurea; woods near, Anemone apennina; south end of common, Littorella lacustris. This common is an undulating, turfy, furze-grown slope, with a gravelly subsoil, and although of some extent is unproductive from a botanical point of view. The bordering woods are prolific in ordinary woodland plants: near, Alchemilla vulgaris; meadows by the canal, E. of Berkhampstead, Geum rivale (?); woods about, Habenaria chlorantha; Hypericum Androsæmum; Lathræa Squamaria.

73. GERARD'S CROSS AND STOKE COMMONS.

These commons are about six miles west of that section of the Colne which runs between Rickmansworth and Uxbridge. They are open heaths, differing in character from those of Surrey, inasmuch as the subsoil is gravel and not sand. The flora consists of the more ordinary heath plants, with little out of the common. The country west of the Colne at this distance is of a hilly aspect, covered to a considerable extent with plantations and woodlands nearly all enclosed. Subsoil, gravel; underlying stratum, chalk. The Furze upon the heaths is mostly of the smaller species; and Heath proper is scarce, Ling predominating. Stoke Common is a broad level expanse, with patches only of moory ground.

*Aira flexuosa.
*Calluna vulgaris.
*Carex cæspitosa; *binervis (abundant).
*Erica cinerea; *Tetralix.
*Juncus squarrosus.
Linaria repens (hedge near the Sefton Arms, Stoke).
*Molinia cærulea.
*Nardus stricta.
*Pteris aquilina.
*Ulex Gallii (mainly); europæus.

Gerard's Cross Common, one mile in its longest diameter from N.W. to S.E. Of greater extent formerly, now much curtailed, and in part enclosed. It slopes to the south-westward into a marshy hollow, bordered by a dark wood.

*Aira flexuosa.
*Calluna vulgaris.
*Carex cæspitosa; *binervis (abundant); *panicea; flava.
*Erica cinerea; *Tetralix.
*Hydrocotyle vulgaris.
*Juncus squarrosus.
*Molinia cærulea.
*Nardus stricta.
*Pteris aquilina.
*Scutellaria minor.
*Ulex Gallii (mainly); *europæus.

Iver Heath is no longer in existence, except nominally. It has long since been enclosed.

74. BURNHAM BEECHES AND FARNHAM COMMON, WITH ADJOINING WOODS.

This locality, in the neighbourhood of those described in the preceding section, is nearer Stoke than Gerard's Cross, and about four miles north of Slough; together they cover a considerable expanse of ground sloping

N

to the westward, and separated from each other by a boggy hollow draining into some ponds below. The common, a mile in length, formerly much more extensive, stretches along the southern border of the wood. The Beeches are of large size; and the place is a favourite and well-known resort of holiday makers. Subsoil, gravel; understratum, chalk; much boggy and marly accumulations in the bottoms. Besides Beech; Oak, Birch, Maple, Holly, Sallow, Viburnum Opulus, and Hazel, are frequent, with an undergrowth generally of Bracken; Ling on the common very dense.

*Aira cæspitosa; flexuosa (common).
*Anagallis tenella (bog).
*Angelica sylvestris.
*Betonica officinalis.
Campanula Trachelium.
*Carex binervis (abundant on the common); *stellulata; *ovalis; *cæspitosa; *flava; *panicea; *vulgaris.
Carlina vulgaris.
*Digitalis purpurea.
*Drosera rotundifolia; intermedia.
*Erica Tetralix; cinerea.
*Eriophorum angustifolium.
*Euonymus europæus.
*Euphorbia amygdaloides.
*Genista anglica.
*Hieracium umbellatum; *vulgatum.
*Hydrocotyle vulgaris.
*Hypericum hirsutum; *humifusum; *perforatum; *Elodes; *pulchrum.
*Juncus squarrosus; obtusifolius, &c.
*Luzula sylvatica.
*Lysimachia nemoralis; *Nummularia.
Malaxis paludosa
*Melampyrum pratense.
*Menyanthes trifoliata.
*Molinia cærulea.
*Nardus stricta.
*Orchis maculata.
*Oxalis Acetosella.
*Potamogeton polygonifolius.
*Primula vulgaris.
*Rhynchospora alba.
Rosa rubiginosa; *micrantha.
*Rubus glandulosus.
*Scabiosa succisa.
*Scirpus multicaulis.
*Scutellaria minor.
*Senecio sylvaticus.
*Serratula tinctoria.
*Solidago Virgaurea.
*Teucrium Scorodonia.
*Ulex Gallii; europæus.
Utricularia intermedia (ponds).

CRYPTOGAMS.

*Aspidium aculeatum and others (?)
Asplenium Adiantum-nigrum; Trichomanes.
Athyrium Filix-fœmina.
*Aulocomnion palustre.
Equisetum sylvaticum.
*Fontinalis antipyretica (in the rills).
*Hypnum cuspidatum; *splendens and other woodland varieties in plenty.
Leptodon Smithii, &c.
*Leucobryum glaucum.
*Lomaria Spicant.
*Nephrodium Filix-mas, with
Osmunda regalis (?), formerly plentiful. "The king fern has not been found here for years; too many people have been after it." (Local information.)
*Sphagnum cymbifolium; *acutifolium.

75. THAMES DISTRICT, ABOVE AND ABOUT WINDSOR.

Acorus Calamus (Staines Common).
Actinocarpus Damasonium (Winkfield Plain and Bracknell near Windsor).
Bromus erectus (Bisham Wood and Winter Hill).
Calamagrostis Epigejos (about Virginia Water).
Carex lævigata (Windsor Great Park).
Convallaria majalis (Clifden Wood).
Cuscuta europæa (about Reading and Maidenhead); Trifolii (Winter Hill).
Daphne Mezereum (Bisham Wood).
Elatine hexandra (Dam Head Cascade, Virginia Water).
Epipactis latifolia (Bisham Wood).
Fritillaria Meleagris (meadows about Maidenhead and Reading).
Hypericum montanum (Bisham Wood).
Iris fœtidissima (Bisham Wood).
Leucojum æstivum (about Windsor).
Limnanthemum nymphæoides (in the Thames at Cookham).
Littorella lacustris (Windsor Park).
Myosurus minimus (cornfields about Slough, and at Cookham).
Neottia Nidus-avis (Bisham and Clifden woods).
Nymphæa alba (Staines; Windsor).

Œnanthe silaifolia (meadows, Eton); Phellandrium (ponds, foot of Winter Hill).

Ophrys apifera (Cookham and Bisham woods).

Orchis militaris (downs near Reading, both sides of the river, also in Bisham Wood).

Ornithogalum umbellatum (meadows foot of Winter Hill).

Polemonium cæruleum (Windsor, near Reading).

Ranunculus trichophyllus (ponds, foot of Winter Hill); Lingua (Cookham, by the Strand).

Rhynchospora alba (Windsor Park).

Sagina nodosa (Cookham).

Sanguisorba officinalis (Sonning meadows)

Sium latifolium (Sonning, and foot of Winter Hill).

Utricularia vulgaris (ponds, foot of Cookham downs).

THE END.

LONDON: PRINTED BY WILLIAM CLOWES AND SONS, STAMFORD STREET AND CHARING CROSS.

www.ingramcontent.com/pod-product-compliance
Lightning Source LLC
Chambersburg PA
CBHW021708210326
41599CB00013B/1572
* 9 7 8 3 3 3 7 1 2 8 0 5 0 *